PERGAMON INTERNATIONA
of Science, Technology, Engineering an... ...d Social Studies

The 1000-volume original paperback in aid of education,
industrial training and the enjoyment of leisure

Publisher: Robert Maxwell, M.C.

AN INTRO... ...N TO
FEEDING ... LIVESTOCK

...E PERGAMON TEXTBOOK
...SPECTION COPY SER...

The inspection copy of any book published in the...
will gladly be sent to academic staff without o...
course adoption or recommendation. Copies...
days from receipt and returned if not suitable...
recommended for adoption for class use and tl...
12 or more copies, the inspection copy may be reta...
Publishers will be ...sed to receive suggestions fo...
to be published in ...important International Lib...

D1333566

Other Titles in the Pergamon International Library

DILLON, J. L.
The Analysis of Response in Crop and Livestock Production, 2nd edition

DODSWORTH, T. L.
Beef Production

GARRETT, S. D.
Soil Fungi and Soil Fertility

GILCHRIST SHIRLAW, D. W.
A Practical Course in Agricultural Chemistry

HILL, N. B.
Introduction to Economics for Students of Agriculture

LAWRIE, R. A.
Meat Science, 3rd edition

LOCKHART, J. A. R. & WISEMAN, A. J. L.
Introduction to Crop Husbandry, 4th edition

MILLER, R. & MILLER, A.
Successful Farm Management

PARKER, W. H.
Health and Disease in Farm Animals, 2nd edition

PRESTON, T. R. & WILLIS, M. B.
Intensive Beef Production, 2nd edition

ROSE, C. W.
Agricultural Physics

SHIPPEN, J. M. & TURNER, J. C.
Basic Farm Machinery, 2nd edition

YEATES, N. T. M., EDEY, T. N. & HILL, M. K.
Animal Science: Reproduction, Climate, Meat, Wool

AN INTRODUCTION TO FEEDING FARM LIVESTOCK

by

ROBERT H. NELSON, B.Sc.

Second Edition

PERGAMON PRESS

OXFORD · NEW YORK · TORONTO · SYDNEY · PARIS · FRANKFURT

U.K.	Pergamon Press Ltd., Headington Hill Hall, Oxford OX3 0BW, England
U.S.A.	Pergamon Press Inc., Maxwell House, Fairview Park, Elmsford, New York 10523, U.S.A.
CANADA	Pergamon of Canada, Suite 104, 150 Consumers Road, Willowdale, Ontario, M2J 1P9, Canada
AUSTRALIA	Pergamon Press (Aust.) Pty. Ltd., P.O. Box 544, Potts Point, N.S.W. 2011, Australia
FRANCE	Pergamon Press SARL, 24 rue des Ecoles, 75240 Paris, Cedex 05, France
FEDERAL REPUBLIC OF GERMANY	Pergamon Press GmbH, 6242 Kronberg-Taunus, Pferdstrasse 1, Federal Republic of Germany

First edition 1962
Second edition 1979

British Library Cataloguing in Publication Data
Nelson, Robert Howard
An introduction to feeding farm livestock.—
2nd ed.—(Pergamon international library).
1. Animal nutrition
I. Title
636.08′4　　SF95　　78–40839

ISBN 0–08–023757–6 (Hardcover)
ISBN 0–08–023756–8 (Flexicover)

Printed in Great Britain by Fakenham Press Limited, Fakenham, Norfolk

Contents

Preface to the Second Edition   vii

Preface to the First Edition   ix

PART I   The Animal and Its Food

Chapter 1   Introduction—the functions of food   3

Chapter 2   The constituents of food: water   9

Chapter 3   The constituents of the dry matter: carbohydrates   13

Chapter 4   The constituents of the dry matter: fats and oils   19

Chapter 5   The constituents of the dry matter: proteins   23

Chapter 6   The constituents of the dry matter: minerals   29

Chapter 7   The constituents of the dry matter: vitamins   34

Chapter 8   Digestion and the digestive system of farm animals   38

Chapter 9   Feeds commonly used in livestock feeding   49

PART II   Rationing Farm Livestock

Chapter 10   Terms used in animal nutrition   73

Chapter 11   Feeding dairy cattle for milk production   83

Chapter 12   Rations for beef cattle   102

Chapter 13   Rations for sheep   118

Chapter 14   Rations for pigs   138

Index   153

Preface to the Second Edition

THE basic objectives of the second edition remain the same as for the first edition—to provide a concise text in as plain a language as possible on the nutrition and rationing of farm livestock. Students attending part-time and full-time courses at the advanced craft/technician level will find the book a useful supplement to lectures, but the step-by-step approach will appeal to anyone requiring the fundamentals of feeding farm livestock.

In the period between the two editions the relative importance and availability of some farm feeds has changed. The industry has also become metricated; but, most important of all, a new basis for the scientific formulation of rations for ruminant stock has been developed—the metabolisable energy system. All these features have been incorporated in the new edition with the chapters on rationing stock having been completely re-written.

When dealing with the rationing of the various classes of stock, extracts or simplified versions of tables of feed analysis and feeding standards have been used to illustrate the points under discussion. These have been taken from various publications, but mainly from that of the Scottish Agricultural College's publication No. 29, *Nutritional Allowances for Cattle and Sheep*, which is strongly recommended for anyone requiring a comprehensive set of tables including mineral and vitamin requirements.

Edinburgh, 1978

R.H.N.

Preface to the First Edition

IN WRITING this book I have tried to keep before me the needs of the young man or woman attending a course of Technical Training in Agriculture—either on a part-time or full-time basis. I well realise the difficulty that many students have of settling down to study in an evening after rising early and following a long day of physical work in the open air. For this reason the chapters in the following pages are short and concise and each one should not take more than 5 to 15 minutes to read.

The text covers the livestock feeding in the following syllabuses:

City & Guilds Animal Husbandry (266)
U.L.C.I. Animal Husbandry
U.E.I. Foods and Feeding

Part-time students following the above courses should use the book in conjunction with the lectures. Full-time Farm Institute students will also find it a supplement to the livestock feeding aspects of the Animal Husbandry lectures.

Many students, while understanding the content of a chapter or lecture, have difficulty in "giving the material back" in written or oral work. The questions at the end of each chapter are designed to help in this respect, and it is a useful exercise to write out each set of answers in full, with or without reference to the text. If the questions can be answered correctly without reference to the text, then the main points of the subject will have been understood.

The "practical work" is also designed to illustrate the material covered in each chapter and to relate the teaching to everyday practical work on the farm.

It is sincerely hoped that by the four media—being taught in the classroom, reading the material, answering the questions and doing the practical exercises—many students will be able to understand and make progress with this relatively difficult subject.

Newton Rigg, 1962   R.H.N.

PART 1

The Animal and its Food

CHAPTER 1

Introduction—the functions of food

GENERAL

LIVESTOCK farming is basically concerned with taking crops and their byproducts and converting them through animals into products suitable for human consumption. Crops such as grass, oats and root crops, which are largely unsuitable for human consumption, together with byproducts such as sugar-beet tops, oil cakes and many others, are fed to livestock to produce milk, meat, and eggs, which are edible by man. Livestock feeding, is therefore, a most important part of livestock farming, and most other aspects of livestock husbandry are directed to making this process more efficient.

Feeding is also the most expensive item in the production costs of livestock farming. The bill for the purchase of feedingstuffs is a major expense on most farms, and on those farms that endeavour to be more self-sufficient, a great deal of the farmer's time and energy is devoted to growing more and better crops for feeding to his stock. Here the expense of feeding is spread over other items such as fertiliser, fuel and machinery costs.

With some types of stock it is easy to see that food is the major expense, particularly where much of it is in a concentrate form, and has to be bought in. In pig feeding at least 80% of the costs are feed costs, and when a tonne of pig meal is delivered to a farm, twenty pigs nearing bacon weight will consume it in 3 weeks or less—and the expense can be directly seen. With other classes of stock, particularly grazing animals, the food costs are often masked, as no direct payment is made for much of the food they consume, but the expense still remains. When the cost of producing milk is analysed, some two-thirds of the cost is food fed to the cow. Approximately 60% of the feed cost is

3

spent on purchased feeds and 40% on homegrown feeds; it is fairly easy to calculate the cost of the purchased feeds but the amount spent on the home-grown feeds is more difficult to determine.

This is yet another reason why a study of the principles of feeding is so important: because food costs are so high, wasteful feeding can turn a profit into a loss—or, on the other hand, by giving careful thought to the rationing, a producer has the greatest chance of improving his profits.

Finally, by better mechanisation, improved varieties of crops, and a greater use of fertilisers, it has become possible to obtain greater crop yields. In many cases there is now the problem of finding the most economical way of utilising the increased yields. For instance, many farmers are growing greater yields of better quality grass, but a great deal of thought is required on how to use this increase efficiently. In this respect there is a strong link between stock farming and crop farming, and any successful livestock farmer must have a sound knowledge of crop production. This also illustrates the necessity for skill in feeding and management by the **individual farmer** on his **particular farm** with his **particular stock**, an aspect of feeding which cannot be learnt from a text-book, but must be gained by considerable experience on the farm. What can be learned, however, are the general guiding principles and the underlying science related to the practice, and that is the purpose of these chapters.

THE FUNCTIONS OF FOOD

The word "food" can only be described in very general terms, such as "material which gives nourishment to the body" or "fuel for the body". Neither of these definitions is really adequate, and it is probably better to consider what food does rather than try to define it precisely.

The functions of food are as follows:

(a) To supply the body with heat and energy.

Just as an engine requires a regular supply of fuel to give it power, so the body requires a regular supply of its own particular fuel. Even a relatively inactive animal is using considerable quantities of energy. Its heart pumps hundreds of litres of blood each day and its intestines squeeze several hundred kilograms of food and water through its body. Animals are also losing heat from their bodies—a sow, for example, gives off as much heat as two infra-red lamps. This heat has to be replaced as the body temperature must remain constant if it is to work properly.

(b) To supply nutrients for the repair of worn-out tissues.

The cells of the body are continually dying and being replaced by new ones. The outer layers of the skin are composed of layers of dead cells, which are rubbed off and replaced by others from the lower layers. This process of repair and replacement is a continual one, and applies to all body tissues.

(c) To supply nutrients for the body to increase in size and weight.

Most farm animals are growing. When fully grown they are either slaughtered for meat, or kept on for breeding, or for some other form of production. The extra nutrients necessary for the increase in weight must come from the food.

(d) To supply nutrients for the production of offspring.

Early in the pregnancy period the developing foetus (young) does not take much nutrient from the dam's body. In the later stages it grows rapidly and increased feeding is essential. Laying birds also can be considered in this category.

(e) Finally, food supplies nutrients to the body for some form of production, such as milk, work (in the case of the horse), and wool.

Milk is a rich food containing considerable quantities of nutrients for the benefit of the young. These nutrients must be replaced through the food, otherwise the animal drains them from its own body and loses condition.

To summarise—food is the material that provides the body with nutrients. The functions of these nutrients are:

(a) to provide heat to maintain body temperature;
(b) to provide energy for the essential processes;
(c) to repair worn-out tissues;

(d) for growth and increase in weight;
(e) for the production of offspring;
(f) for the production of milk (or other products).

MAINTENANCE AND PRODUCTION

The functions listed above can be divided into two categories. The first three, (a), (b), and (c), are said to be for **maintenance** and the last three, (d), (e), and (f), are for **production**.

In calculating an animal's feed requirements it is practically convenient to make an allowance for maintenance and a further allowance for production. These two together give the total nutritional requirements of the animal, and a daily ration can be devised to meet this need.

The larger the animal the more food it needs to maintain itself. For example, the maintenance needs of a Jersey cow could be met from a daily ration of 5 kg of hay or 3.5 kg of barley whereas a large Friesian cow would need 8–9 kg hay or 5 kg barley. Similarly with production— the amount of feed has to be increased as production increases. A cow giving 30 kg milk per day requires more food than one giving 20 kg. With meat-producing animals the rate of growth can be controlled to some extent by increasing the feed to obtain more production in terms of daily liveweight gain.

It must never be thought, however, that an animal uses particular feeds in its ration for maintenance and other feeds for production. Such an impression could easily arise from the fact that, for example, as a dairy cow's yield increases and she has a need to take in more nutrients and it is usual to give her more concentrate feeds, whereas when she is giving very little or no milk she is fed mainly bulky farm foods. Production becomes linked with concentrates and maintenance with hay or silage. As far as the animal is concerned, both concentrates and the bulk feeds contribute equally to both maintenance and production, and what is important is that the cow's **total nutritional requirements** are met.

This is the basis of the recently introduced methods of rationing stock, and the maintenance and production allowances are simply a convenient means for calculating an animal's total requirements. The final ration will doubtless consist of a variety of feeds, some home-grown bulk feeds, some home-grown or purchased concentrates, which combined together will meet the animal's needs for both maintenance and production.

THE RELATIONSHIP BETWEEN THE ANIMAL AND ITS FOOD

The grazing animal is able to convert the grass it eats into either meat or milk. Obviously there must be some relationship between the grass, the cow that eats it, and the milk it produces. This is in fact the case, for the animal, its food and what it produces, can be analysed and shown to consist of the same basic constituents.

A major constituent of the animal and its food is **water**. The rest is termed **dry matter**, which is the material left after the water has been removed (dehydrated).

There are five major constituents of the dry matter:

(a) **carbohydrates**, i.e. starches and sugars;
(b) **fats** and fat-like substances;
(c) **proteins**;
(d) **minerals**;
(e) **vitamins**.

The constituents of the dry matter give food its feeding value as there is no feeding value in the water content. Silage with a high dry matter content has a better feeding value than silage with low dry matter providing each has been equally well made from similar material. The actual **composition** of the dry matter is also important in determining the final feeding value. Poorly made hay from a poor quality meadow can have the same dry matter content as in-barn dried hay made from first-class ley, but because of the difference in the constituents of the dry matter the latter has the greater feed value.

The constituents of the dry matter are not single substances, but groups of substances, the members within each group having similar chemical composition. The main group in plants are the carbohydrates, whereas in animals the proteins are important. In both animals and plants, minerals and vitamins occur in relatively small amounts. Any farm feed can be analysed to determine the proportions of the different constituents present. Alternatively, it is possible to refer to feed analysis tables, which give an analysis of all in common use. Such tables can be found in the HMSO Technical Bulletin No. 33.

The following chapters deal with each of the **constituents of food** in more detail.

Having read this chapter, you should be able to **answer these questions:**
1. What is food?
2. What are the six functions of food in the body?
3. What is a maintenance ration?
4. What is a production ration?
5. What are the five constituents of the dry matter?
6. What is the relationship between the dry-matter content and the feeding value of a food?
7. Are carbohydrates characteristic of plants or animals?
8. Which constituents only occur in relatively small quantities?

CHAPTER 2

The constituents of food: water

WATER is essential for all classes of livestock. Whereas an animal deprived of food can live for some considerable time, without water it will die in a few days. Some classes of stock can live without **drinking** water, but this is only when their food is extremely succulent. For example, sheep can be folded on roots in areas such as the Yorkshire Wolds, since this class of stock can exist with no water apart from that contained in the roots they are grazing.

Milking animals, on the other hand, require a large amount of water, and this should always be available to them. A dairy cow, when not in milk, requires 30–40 litres per day. A further 4 litres per day is required for every litre she produces when in milk. Similarly, lactating ewes and sows should always have water available, or otherwise the milk yield will quickly fall. Alternatively, when an animal is to be dried off, it is necessary to restrict the water as well as the food intake. An animal requires more water under hot, dry conditions, than when under cooler, wet conditions when evaporation from the body is appreciably less. More water will be drunk if an animal is on a dry fibrous ration, i.e. the food has a higher dry matter content, than when fed on succulents with a low dry matter. Providing the feeding is correct, there is no class of farm animal that does not benefit from having water always available. The only possible exceptions may be fattening pigs and calves under certain rearing systems.

WATER IN THE ANIMAL

Fifty to seventy per cent of the weight of an animal's body is water.

In general, as an animal gets older, the water content becomes proportionately less. In a calf weighing 50 kg the water content may be 70% whereas when the same animal reaches 600 kg it falls to 50%. Pure fat contains no free water; therefore a fat animal contains proportionately less water than a lean one. This applies to all classes of stock. (These figures assume the animal has no water or food in its digestive system.)

Water is therefore an essential component of animals, and it has some very important functions in the body, which are as follows:

(a) It acts as a solvent, i.e. it allows other substances to be dissolved in it, and by this means they can be transported round the body.

(b) It acts as a medium in which chemical reaction can take place—chemical reactions are going on in the body continuously, and these usually take place between substances that are in a state of solution in the body fluids.

(c) It removes surplus heat by evaporation from the surface of the body—perspiration or sweating. A considerable amount of heat is required to evaporate small amounts of water, hence the body can lose heat quickly if the surface is moistened by perspiration. (*Water has a high latent heat of vaporisation.*)

(d) It helps to maintain a steady body temperature—this is because water reacts slowly to a sudden change of temperature. (*Water has a high specific heat.*) It is also a means by which heat can be circulated round the body.

(e) It gives the body shape, i.e. if the body cells lost their water they would give way like a balloon losing its air—the tissues would then lose their resilience and the body lose its shape. (*The water is providing a mechanical means of maintaining the turgor of the cells.*)

There is never any **pure water** in an animal as it is always in combination with other substances or has substances dissolved in it.

WATER IN THE FOOD

The amount of water in the food varies a great deal. It is on the water content that foods are largely classified. The succulent foods are those which contain a large amount of water. The concentrated foods contain relatively little water—they have a concentration of feeding value due to their high dry matter content. The percentage variation in water content is illustrated below:

Succulents—roots, tubers, and green fodders	80–90
Silage	75–80
Cereal grains, milling offals, peas, and beans	12–16
Concentrates of animal origin	10–15

In general, fresh young material has a higher water content than older, mature crops. Any crops which are dried for storage have to be brought down to 15–20% moisture or less. Cereal grains can be stored in bulk at 14–16% moisture. With other concentrates, such as bought in cakes, the moisture content largely depends on the processes to which they have been previously subjected.

For any feed it is an easy matter to calculate the water content by subtracting the percentage dry matter from 100. In practice, however, it is the dry-matter content in which the feeder is primarily interested. The tables give the dry matter as so many **grams per kilogram (g/kg)** of the feed; e.g. tower silage (grass), 400-g/kg.

Dividing the 400 by 10 gives the percentage dry matter, i.e. 40%, and from this, if required, the percentage of water: 100–40 = 60%.

QUESTIONS

1. Name a class of stock that may be kept without water to drink, and under what conditions.
2. How much water does an average size dairy cow require—when dry —when in milk?
3. Which animal body contains most water—a fat one or a leaner one— an old one or a young one?
4. State briefly five functions of water in the body.

PRACTICAL

1. From a feed analysis table, calculate the water content of: marrow stem kale (unthinned), good quality seeds, hay, barley (grain), linseed cake, white fish meal.

2. What is the variation in the dry-matter content of the various types of mangold?
3. How do mangolds compare with fodder beet in this respect?
4. What is the dry-matter content of your silage this season?
5. What was the moisture content of the grain as it came off the combine this year?
6. What is the moisture content of the grain as it now stands in storage?

CHAPTER 3

The constituents of the dry matter: carbohydrates

CARBOHYDRATES are characteristic of plants. That is to say, a large proportion of the dry matter of plants, and feeds derived from plants consists of carbohydrates in one form or another. The entire skeletal structure of green plants is made from this type of material. With animals the picture is very different. Carbohydrates are only present in very small amounts in the animal body because they are energy foods, which after digestion the animal uses immediately as a source of heat and energy.

THE CHEMISTRY OF CARBOHYDRATES

All carbohydrates consist of three elements only—carbon (C), oxygen (O), and hydrogen (H). These three can be combined in many ways, however, each combination resulting in a different substance with different chemical and physical characteristics. The chemists group them into three general groups according to the complexity of their chemical make-up.

1. **Monosaccharides.**—The monosaccharides are the simplest sugars, one of which is **glucose**. This is one of the first carbohydrates to be formed in the plant in the process of **photosynthesis** (see later). It is also the principal substance into which the more complex carbohydrates are broken down when they are digested by the animal. Should an animal require an immediate supply of energy, glucose can be given in a drench, or injected into the bloodstream.

2. **Disaccharides.**—The disaccharides are more complex sugars;

13

sucrose, the common sugar used in the home, is a typical example; **lactose** or milk sugar, and **maltose** or malt sugar are others. They are all white crystalline compounds with a sweet taste and, with the exception of lactose, are manufactured in **plants** from the simpler monosaccharides.

3. **Polysaccharides.** The polysaccharide group is the most complex. Although these substances are built up from the simpler carbohydrates they bear little physical resemblance to them. **Starch**, which is the main energy reserve in plants, is an important member of this group.

Another important polysaccharide is **cellulose**, the main material in the cell walls. Cellulose fibres can be digested by ruminants and make up the important **digestible fibre** part of the feed for cattle and sheep.

As plant cells grow older the cellulose is replaced by a rough substance called **lignin**. It is largely indigestible even by ruminants and, therefore, contributes nothing to the feed value.

It can be seen, therefore, that although the carbohydrates are composed only of three elements; they range from simple, easily digestible sugars to complex and often indigestible fibres.

Carbohydrates in the plant and in the food

Glucose is manufactured in the leaves of plants by the process of **photosynthesis. Carbon dioxide** is taken from the atmosphere and combined with water taken in by the roots. This process can be carried out only by the **living plant** in the presence of **chlorophyll** (the green colouring matter in leaves) and **sunlight.** The chlorophyll acts as a stimulant (a catalyst) to the reaction, and the sunlight provides the necessary **energy**, which is locked up in the process of combining the carbon dioxide and the water. It is the same energy that is released when the sugar is finally broken down in the body. Thus the sun is the ultimate source of energy for all life, and this energy becomes available to the animal via the activities of plants.

The simple sugars first formed are quickly converted into the more complex disaccharides. Some plants store their energy in this form, e.g. sugar-beet stores sugar in its root. An important part of the feeding value of mangolds is due to the sugar they contain. Other plants go one step further and convert the sugar into starch, which has a greater concentration of energy, weight for weight, than sugar. In this way plants are able to store more energy in less space. Potatoes are well known as a starchy food—the starch being stored in the tubers. Many seeds also contain starch, cereal grains in particular.

Cellulose is the end product of the chain of carbohydrate synthesis, and together with the lignin make up the "crude fibre" part of a feed. They are the main substances in the plant skeleton and the proportion of lignin increases as the plant gets older. Cellulose is easily digested by ruminants, but the pig can only cope with a small quantity. Lignin is indigestible, but ruminants need a certain amount in the form of coarse fodders or roughages to keep the stomach contents at the correct physical consistency. The low feeding value of poor stemmy hay is mainly due to a high lignin content.

Most livestock feeds are derived from plants, so the proportion of carbohydrates in them is generally high. The exceptions are those foods, such as fish meal, which are of animal origin where the proportion of carbohydrates is very low, and those foods which are byproducts of a process which has deliberately extracted the carbohydrate previously. An example is bran, which is a byproduct of flour milling. Because flour has been extracted, the carbohydrate content of bran is much lower than that of the original wheat.

In feed analysis tables the amount of carbohydrate is given under "carbohydrates" or "nitrogen-free extractives". The cellulose/lignin content is given under "crude-fibre".

Carbohydrates in the animal

(a) As a source of energy

Carbohydrates are the main source of heat and energy for the farm animal. Whereas the process in the plant was one of **synthesis** (building up), in the animal this is reversed. Whatever the carbohydrate in the diet, the animal endeavours to break it down into simple substances. Starch is converted to sugars and complex sugars into simpler sugars and so on. The final product of carbohydrate digestion is **glucose**, which is absorbed through the wall of the gut into the blood-stream. The end of the process is reached when the glucose is broken down by the body cells and the energy released. This is the process of **respiration** and it is the exact opposite of **photosynthesis.** Oxygen is required for this and is obtained from the lungs via the blood-stream. The byproducts—the original carbon dioxide and the water—are also carried away in the blood-stream back to the lungs where they are expired.

Respiration can be represented as follows:

Glucose + Oxygen = Carbon dioxide + Water + Energy (or heat)
$$C_6H_{12}O_6 + 6O_2 = 6CO_2 + 6H_2O$$

Respiration is going on continuously so the carbohydrate in the animal is quickly used up, and at any one time there is very little present. There is some in the gut during digestion, and glucose occurs in the blood and liver (which acts as a temporary store), and again in the muscles during muscular activity. The milking animal has lactose (milk sugar) in her milk.

Diabetic animals are not able to break down glucose, and to avoid too great a concentration in the blood it is excreted by the kidneys and then appears in the urine.

(b) As a fattening food

. The fat in an animal is largely manufactured from carbohydrates and not, as one might expect, from the fat in the diet. If an animal eats a greater quantity of carbohydrate than it needs for its immediate energy requirements, the surplus is converted and stored as fat. In other words when the energy intake of the animal exceeds its energy output the surplus is stored as fat. Both these factors can be controlled by the feeder. The energy intake can be increased by feeding high energy foods such as oats, barley, flaked maize, and so on. The energy output can be reduced to a minimum if the animals are kept quiet and in warm, draught-free buildings. Under these conditions fattening will proceed satisfactorily. Alternatively, a group of animals such as pigs, kept in cold, wet conditions with too little trough room, will have a high energy output trying to keep warm and fighting for food, and the fattening rate will be correspondingly unsatisfactory.

To summarise—carbohydrates are characteristic of plants and are found, therefore, in considerable quantities in the majority of feeds. They are utilised by the animal for heat and energy and for converting into fat. Animals need a supply of carbohydrate in the diet at all ages, and particularly those animals that are being finished to go to the butcher. It must also be remembered that not all farm animals can digest fibre, but a certain amount is necessary for the ruminant.

QUESTIONS

1. What is meant when it is said that carbohydrates are characteristic of plants?
2. Why is so little carbohydrate found in the animal?
3. Name the three chemical elements in carbohydrates.
4. What name do we give to the simplest sugars? Name one of these sugars.
5. Name three disaccharides.
6. Which polysaccharide forms the main energy reserve in plants?
7. Which two substances make up the crude fibre?
8. Which of the two is digestible, particularly by ruminants?
9. Why do ruminants need the other one in the diet?
10. What is the process by which carbon dioxide is combined with water to make sugar?
11. Where does this process take place, and under what conditions?
12. Name two farm crops that store sugar in their roots.
13. Name two farm crops that store starch—in which part of the plant do they do this?
14. Although most feeds have a high percentage of carbohydrate, can you name a group where the percentage is low?
15. What do you understand by the term respiration?
16. Where is carbohydrate stored temporarily in the body?
17. What is lactose?
18. What feeds can be fed to give an animal a high energy intake?
19. How can you reduce the energy output of an animal?

PRACTICAL

1. From feed analysis tables:
   (a) List the figures for digestible carbohydrates (nitrogen-free extractives) and crude fibre for the cereals, barley, and oats. Which of these would you say had the greatest fattening value weight for weight?
   (b) Compare the crude fibre in hay from grass (high digestibility) with hay from grass (low digestibility). The crude fibre in the first is 70% digestible but in the second only 56% digestible. What is the reason for this difference? Which one was made early in the season?
2. If you grow sugar-beet, find out the sugar content for this year's consignments. If these vary can you account for this variation?
3. If you fatten pigs, lambs, or bullocks, list the items in the concentrate mixtures for these stock. Point out which are the high energy fattening feeds, i.e. those that have a high proportion of carbohydrate.

CHAPTER 4

The constituents of the dry matter: fats and oils

FATS, oils, and closely related substances such as waxes are widely distributed in both plants and animals. In animals, fat is the main energy reserve, and in the wild state an animal builds up its fatty reserves during summer and draws on them during winter when food is scarce. In plants, oil is often found in the seeds, sometimes in quite considerable amounts, so much so in fact, that it becomes worth extracting; it is then used for a variety of purposes, such as in the manufacture of soap and margarine.

THE CHEMISTRY OF FATS AND OILS

Fats and oils, waxes, and other substances are grouped together mainly because of their similar physical characteristics. As a group, the chemist knows them as the **lipids**. The fats and oils are by far the most important members of the group and for the purpose of this study the rest may be ignored. Nevertheless, all have this in common in that they are insoluble in water, but they are soluble in fat solvents such as ether, chloroform, alcohol, and carbon tetrachloride (these substances are sometimes used for removing fatty stains from clothing).

The fats and oils—like the carbohydrates—consist of the three elements—carbon, hydrogen, and oxygen. Any single fat or oil is built up from simpler substances. One of these is **glycerol** and the others are known as **fatty acids**. When an animal digests fat or oil it breaks it down into glycerol and the particular fatty acids it contains.

Although there are several fatty acids there are three which occur in greatest quantities. They are:

19

   (i) stearic acid;
  (ii) palmitic acid;
 (iii) oleic acid.

The first two are said to be "saturated", which is a chemical term and it is not necessary to understand its full meaning. The third one, oleic acid, is termed "unsaturated". The important point is that the "saturated" pair have a different effect on the fat from the "unsaturated" acid.

Stearic and palmitic fatty acids tend to make a *fat* that is solid at normal temperatures. Oleic fatty acid tends to make an *oil* at normal temperatures. This is the basic difference between fat and oils. Fats are composed largely of stearic and palmitic fatty acids and oils of oleic fatty acid. The significance of this will be seen shortly.

Fats and oils in the plant and feeds

Fats and oils are found in most plants. The seed often contains the greatest amounts of oil; it also occurs in stems and leaves but there is very little in the roots. This is reflected in the analysis of feeds. Concentrate feeds—many of which are derived directly or indirectly from the seeds of plants contain up to 10% oil. Oil cakes, which are the material left after oil has been extracted, still contain between 0.5 and 5.0% oil. The exact amount rather depends on the method of extraction —some methods being more efficient than others.

Hay and similar dried green crops also contain a small amount of oil. It is only in roots such as mangolds, swedes and potatoes that the oil drops to an insignificant amount.

As one would expect, the feeds derived from animal sources contain the most oil and fat. Meat and bone meal feels greasy to handle and may contain 12% fat. The range in the group is from $3\frac{1}{2}$% to 12%. Too much fat of the wrong kind may have harmful effects and it becomes necessary for manufacturers to guarantee that their products only contain a certain percentage. Meat meal for example is marketed with high oil content—11.0% and low oil content, 4.5%. In feed analysis tables the percentage of fat is given under "oil" and "ether extract".

Fat in the animal

Fat in the body is known generally as the adipose tissue. It has two distinct roles:

(a) as part of the structural tissues of the body;

(b) as a reserve of energy.

Tissue fat (a) is always present even in times of starvation. Reserve fat is stored by the animal in times of plenty, being laid down under the skin and in and around various organs of the body. It may also occur between the bands of muscle fibre in fat beasts giving the "marbling" fat. Naturally the amount of reserve fat is very variable, decreasing in times of undernutrition and increasing when food is plentiful. Of all the food constituents, fat has the highest energy value; thus an animal can store a far greater reserve of energy as fat than in any other form.

Most of the adipose tissue in an animal is manufactured from the carbohydrate in the diet. Nevertheless, the fat in the diet does influence the body fat. During digestion fats in the food are broken down into glycerol and their respective fatty acids. If these contain a considerable amount of the unsaturated oleic fatty acid, there may be an adverse effect on the body fat; a soft oily fat, which would go quickly rancid, may be produced in the carcass. Feeds which have such an effect, when fed in large quantities, are maize, palm kernel, and other oil cakes. For this reason the rations of pigs in the final stages of fattening should not contain too high a proportion of these foods. Some swill may contain a high proportion of oil. Barley, on the other hand, a carbohydrate food, is noted for its ability to produce a good hard white fat in pigs.

To summarise—fats and oils, though present in most plants and feeds, are often not present in very large amounts. They are important in that they increase the energy value of a diet and are largely used for this purpose by the animal. The fat in the diet does influence the body fat, however, and care should be taken to avoid the over-use of certain oily feeds in fattening rations.

QUESTIONS

1. What are the three chemical elements in fats and oils?
2. Name two substances in which fats will dissolve.

3. What are the substances into which fats are broken down?
4. Name three fatty acids. Which of these tends to give an oily fat?
5. Which group of feeds contain the greatest amount of fat or oil?
6. How does the fat in the diet influence the body fat?
7. What feeds may cause a soft oily fat?

PRACTICAL

1. From feed analysis tables, compare the amount of oil in:
   (a) groundnut cakes (decorticated and undecorticated);
   (b) extracted groundnut cakes.
2. From the tables, compare the oil content of the common cereals, oats, wheat, and barley with the oil content in maize.
3. Check your farm rations to see what part linseed cake plays. Is it fed to any class of stock in particular, and for what purpose?
4. Obtain a bacon carcass inspection form, and note if provision is made for indicating that soft fat may be present.

CHAPTER 5

The constituents of the dry matter: proteins

ABOUT one-fifth of the animal body is composed of proteins; there is very little carbohydrate, and the rest is mainly fat, minerals, and water. Just as carbohydrates are said to be characteristic of plants, proteins are characteristic of animals. It is not surprising, therefore, that animals require a considerable amount of protein in their diet, and how to supply this food constituent most economically is a major consideration on many farms. Plants, and therefore farm crops, contain only a small proportion, and it is difficult, if not impossible, for many farms to grow enough fodder to supply enough protein for the stock being carried. The more intensive the stocking, the greater the problem.

The only answer to this problem is to buy in feeds containing a high percentage of protein to supplement those grown on the farm. Some of these feeds are byproducts such as the oil cakes and meals, the original material coming from plants. Other protein feeds are of animal origin such as white fish meal or meat-and-bone meal, and these are of particular value in animal feeding, as will be seen later.

CHEMISTRY OF PROTEIN

Proteins are a complex group of substances. Like the fats and the carbohydrates, they are composed of the elements carbon, hydrogen, and oxygen, but in addition they also contain **nitrogen**. This is an important distinguishing factor. Some may also contain small amounts of other elements—**phosphorus** and **sulphur**, but nitrogen is always present. What is more, the amount of nitrogen in the different food proteins is fairly constant—about 16%. When a chemist has to analyse a feed to

23

determine the amount of protein present, he uses this knowledge and actually determines the amount of nitrogen present. It is then an easy matter to calculate the amount of protein by multiplying the amount of nitrogen by 6.25. The figure finally obtained is known as the **crude protein figure**, which forms the basis for a calculation of the protein feeding value of a food. It is the figure usually quoted on the bags of commercial compound feeds.

Proteins are built up from a number of simpler substances called **amino acids.** There are about twenty-five of these, and any one protein is composed of a number of them. Obviously the number of ways of combining twenty-five substances is enormous, which helps to explain the great number of different forms that protein takes. Thus proteins are found in muscle, hair, tendons, and nerves—in fact in most body tissues including the skeleton.

Proteins in feeds

All farm feeds contain some protein but the amount varies considerably from feed to feed. The **succulent feeds** and **green crops** such as turnips, cabbage, and kale contain very little—not more than 1% or 2%, and the **coarse fodders,** oat and barley straw, are also in this category. Generally speaking, an animal cannot obtain all its protein requirements by eating these foods alone unless it is only being kept in store condition and it is able to consume large amounts.

Grazing animals can, however, obtain sufficient protein from **grass**, either fresh or as **hay or silage.** Pasture grass contains something over 2% protein, but this can vary, and first quality grazing may contain twice this amount. Likewise the protein content of hay and silage can be even more variable depending on the crop from which it was made, the stage at which it was cut, and how well it was conserved as hay or silage. The protein in hay can range from 3% up to 8% or even higher when artificial drying is practised. With silage the figures are from $1\frac{1}{2}$% to 3%. A high proportion of **legumes** (clovers) tends to raise this figure.

Home-grown concentrates naturally contain a greater percentage, from a little over 7% in oats and barley, to nearly 10% in wheat. The cereals are not usually grown purely for their protein value, however, and beans and peas give a far greater yield of protein per acre. Conditions have to be right for these legume crops, and they are largely confined to the warmer and drier areas. Both contain around 20% protein.

From what has been said it can be seen that the greatest concentration of protein in the plant appears in the seeds. The initial manufacture takes place in the leaves where the elements carbon, hydrogen, and oxygen, found in the sugars, are combined with nitrogen and the other elements taken in from the soil by the roots. That is why there is a high proportion of protein found in **young leafy grass.** As the plant matures, much of the protein is transferred to the reproductive parts, the seed in particular. As a result, the feeding value of the stem and leaf is greatly reduced immediately after flowering, and by the same process the seed obtains a high protein content.

When certain non-protein parts of seeds are removed (extracted) the protein content of the remainder is increased proportionately. For example, oil-bearing seeds such as cotton seed and linseed have the oil extracted and the residues, which are sold as **cotton cake** and **linseed cake**, have a higher protein content and make valuable farm feeds. Another group of feeds of vegetable origin, which are byproducts of an extraction process, are the **wheatfeeds**—bran and the various grades of weatings. These feeds are different factions of the seed coat of wheat left when flour has been extracted. They contain from 10% to 12% protein which is higher than for the whole grain.

All the above feeds are of **vegetable origin**, but there is one other important group that is derived from animals. The main feeds in this group are **fish meal**, which contains the inedible parts of fish, and the various **meat and bone meals**, which are byproducts of slaughter houses. They are extremely valuable for two main reasons: firstly, they are very high in protein; and, secondly, it is protein derived from **animals**, and the importance of this will be explained shortly.

Protein in the animal

As such a high proportion of the animal is one sort of protein or another, it is not surprising that the daily protein requirement for any one animal is relatively large. Protein is required to repair worn-out tissues and build up new ones. The requirements of the young animal are particularly heavy as it is building up muscle tissue or lean meat, which is itself mainly protein. As an animal matures its need for protein is not so great relative to its size, unless it is producing offspring or milk, when the demands increase proportionately.

It will be remembered that proteins are complex substances and are largely insoluble. Before the animal can utilise any protein it must first be broken down into simpler substances—the component amino acids—and this occurs in the intestine. The identity of the original protein is lost, and it passes through the gut wall as a number of amino acids. The body then selects the amino acids it requires and builds up its own particular body protein. The new body tissue usually bears no resemblance to the original protein in the food.

The **function** of the **protein in the food**, therefore, is to supply the animal with a **selection of amino acids** from which it can build up its own body protein.

The question now arises, what happens if the food protein does not contain all the different amino acids the body requires. Fortunately the body is able to manufacture many of the amino acids for itself from other substances containing nitrogen—these are the "**inessential**" **amino acids**. There are, however, about a dozen that it cannot manufacture, and these must be present in the food—these are the "**essential**" **amino acids**—and without any one of them the animal cannot synthesise its own protein and suffers from malnutrition.* It is here that protein feeds of **animal** origin are so important. The proteins in these are very closely related to the protein the animal is trying to synthesise—far more so than some proteins of vegetable origin. They contain the right kind of amino acids in the right quantities, and so they are utilised most efficiently by the animal—they are said to have a **high biological value.**

Practical aspects

From the practical point of view, it is not usually necessary to know the amino-acid content of each food, but it is good husbandry practice

---

*This statement does not apply to ruminants that are able to manufacture all the amino acids through bacterial action in the rumen.

to feed a variety of foods—particularly in concentrate mixtures—in order that the stock may receive a **variety of protein** from which to meet their amino-acid requirements. It is also good practice to include in a concentrate mixture a protein food of animal origin such as white fish meal, particularly where stock are being kept intensively under unnatural conditions.

Protein is costly, and the higher the percentage of protein in a food the greater the cost per tonne. The protein feeds must not, therefore, be fed indiscriminately or wastefully. The protein requirements of a particular class of stock should be known accurately, and they should be met accurately. A knowledge of the protein content of the various feeds is necessary, therefore, in order that these expensive feeds are not fed wastefully. The level of protein in a feed can be obtained from the feed analysis tables. There is, however, more than one way of analysing a feed for protein and these different methods give different results for the same feed. Possibly the easiest method is the one already mentioned by which the **crude protein** figure is obtained. **Digestibility** is also important, and the figure usually quoted as a measure of the protein value of a feed is the **digestible crude protein** (DCP).

(For further explanation, see Chapter 10, "Terms used in animal nutrition".)

QUESTIONS

1. What proportion of the animal body is protein?
2. Why is it difficult for an intensive stock farm to grow sufficient protein to meet the needs of the stock?
3. How is the deficiency usually met?
4. Apart from carbon, hydrogen, and oxygen, which other element is always present in protein?
5. Name two other elements that may be present?
6. What do you understand by the term "crude protein"?
7. Proteins are composed of a number of simpler substances. What are these called?
8. Which group of farm feeds contain the least protein?
9. Which crops could be grown to obtain the highest yield of protein per acre?

10. What is the approximate protein content of the cereals?
11. Which bought-in feeds contain the highest percentage of protein?
12. To what use does the animal put protein?
13. What happens to protein in the intestine?
14. What is the difference between an "essential" amino acid and an "inessential" amino acid? Does this apply to all animals?
15. What is meant by a protein having a high **biological** value?
16. Why is it desirable to feed a variety of protein feeds in a production mixture?

PRACTICAL WORK

1. List these feeds in order of their digestible crude protein content, and state the current price per tonne of each: (a) white fish meal; (b) weatings; (c) linseed cake; (d) oats; (e) barley; (f) wheat; (g) decorticated cotton cake; (h) groundnut cake; (i) soya bean meal; (j) flaked maize.
2. Have a sample of your hay and/or silage sent for analysis through your local advisory officer, or through the feed representative of the firm with which you deal.
3. If you have hay or silage of good and poor quality, obtain an analysis of each. Does this analysis reflect the conditions under which it was made?
4. If you mix your own rations, list the foods in them in order of their protein content. Work out the protein content for the mixture as a whole (use DCP figures).
5. If an animal consumed 1 kg of this mixture, how much protein would it have eaten?

CHAPTER 6

The constituents of the dry matter: minerals

WHEN an animal or plant is completely incinerated (i.e. burnt), all that remains is **the ash**. This is the **mineral content** of the living organism, and it is obviously only a small part by weight of the whole body. The ash contains the elements silica, potassium, chlorine, magnesium, iron, sodium, phosphorus, calcium, with traces of iodine, manganese, copper, zinc, cobalt, and others.

These elements in the living animal or plant are either in combination with organic substances such as protein, or else they occur as inorganic compounds.* Although small in quantity, minerals are vitally important to the living organism, and besides forming part of the structure, i.e. the skeleton, they also stimulate and control most of the body processes. The process of respiration, for example, may be simply regarded as the breakdown of sugars and release of energy, but it is in fact a whole chain of reactions, and the mineral **phosphorus** plays a very important part at every stage.

MINERALS IN THE PLANT

Plants obtain minerals from the soil in which they are growing. The minerals must be in the form of simple salts dissolved in the soil moisture before they can be taken up by the root hairs into the plant.

If the soil lacks a particular mineral, the plant may suffer,

*An organic substance is one produced by a living organism and invariably contains carbon, e.g. sugars, protein. An inorganic substance is of mineral origins (non-living), e.g. common salt.

consequently, from a **deficiency**. Moreover, such a deficiency may be passed on to the grazing animal which feeds on the plant.

It is not always a case of an absolute deficiency that is the trouble. Sometimes an **imbalance** between minerals occurs. For reasons that are not fully understood, a concentration of one particular mineral may lead to an apparent deficiency of another, which may again show itself as a deficiency in the animal.

In general, for plants of a particular species, the mineral content is fairly constant. There are considerable differences between species, however, and differences also in the different parts of the same plant. The green parts, stems, and leaves, contain a good supply of calcium but relatively less phosphorus. Seeds, on the other hand, contain **more** phosphorus than calcium. Roots contain very little mineral matter at all.

Minerals in feeds

Young leafy material, whether it be grass, kale, or legumes, generally has a **good supply of all the essential minerals** particularly **calcium**. This is also true of the fodders made from them. Good lucerne hay, for example, contains 20 g lime per kilogram.

Cereal grains, peas, beans, and cattle cakes normally have satisfactory amounts of phosphorus and potassium but tend to be poor sources of calcium and chlorine.

Roots and tubers are poor in all minerals. As might be expected, fish meal is rich in calcium, phosphorus, and chlorine, which is another reason why it is a most valuable feed.

Minerals in the animal

The skeleton of an animal is largely mineral matter—calcium and phosphorus in particular. In fact 80% of the minerals in an animal are found in the skeleton. The rest play an important part in the body processes. The stomach, for example, produces considerable quantities of hydrochloric acid during digestion, which necessitates a constant supply of chlorine, and right through the digestive tract minerals are brought into the process of digestion. All the body systems depend on minerals in order to function properly.

When an animal takes in mineral matter in its food, a certain amount —although by no means all—is absorbed into the blood-stream, as are other food constituents. The body has a very sensitive mechanism by which it controls the amount of mineral in its system. Here the kidneys play an important part. They prevent too great a concentration of any one mineral occurring in the blood-stream by "filtering" out any excess and passing it out of the body in the urine.

As in the plant, deficiencies are more often the trouble, rather than excesses. Often these deficiencies occur when the animal is going through some abnormal stress period. The whole food-handling mechanism, or **metabolism*** of the animal as the scientist calls it, is then strained, and if at the same time there is an imbalance or a deficiency in the mineral supply, the animal shows the symptoms of a **deficiency disease**.

There are many examples of this. Young piglets, for instance, grow very rapidly during the first few days of life and can double their birth weight in a week. They rapidly outgrow the supply of iron with which they are born. There is no iron in the mother's milk, and if they are being reared indoors under artificial conditions they have no access to an alternative supply. This alone is sufficient to make the young pigs show the symptoms of **iron deficiency**, i.e. they appear **anaemic**. If the piglets are also subjected to the stress of cold, damp conditions at the same time, these symptoms are far greater. The answer must be to improve the conditions and give the young animals a supply of iron in one form or another.

Another obvious example is **milk fever** in dairy cows. A newly calved high-yielding cow is subjected to the stress of calving together with the sudden production of milk. The metabolism of the animal is disturbed and her **calcium** reserves are drained into the milk she is producing. Through the loss of calcium her nervous system is upset and she thereby loses muscular control. The typical milk fever symptoms of twitching, paddling with the hind feet, are shown, and the animal goes down with exhaustion. Little can be done to remove the stress at this time, although

---

*The metabolism of an animal is the sum total of all the chemical reactions that go on within its living body.

in the past measures were taken to stop the cow producing milk. A supply of calcium, however, administered directly into the blood-stream, can have rapid and quite dramatic results in restoring the animal to normal health.

Practical aspects

Under normal farming systems, where animals are being kept under fairly natural conditions and they have access to good quality grazings, there need be no worry about the supply of minerals. There may be certain areas where the land is short of a particular mineral or trace element, as is sometimes the case with upland grazings, but these are generally known and precautions taken.

Trouble occurs when conditions are no longer normal; in the depth of winter, for example, when only poor quality food is available and stock are housed. Under these conditions mineral deficiencies may occur. Alternatively, the rations may be good, but the system of management is very intensive, the stock being heavily fed to give maximum production. The management and feeding must then be of the highest standard, and a careful watch must be kept on the mineral content of the rations.

Poultry on free range may make up for deficiency in their rations from what they can pick up. If the same birds are placed in battery cages, more eggs may be produced, but every item of their nutrition, including mineral requirements, must be met by the feeder.

Mineral supplements

In practice the feeder guards against deficiencies in **calcium, phosphorus**, and **chlorine**, and mineral supplements that are added to concentrate rations are usually designed to supply these. A simple mixture of **chalk** (to supply calcium), **steamed bone flour** (calcium and phosphorus), and **common salt** (chlorine) may be all that is needed. On the other hand, as only a few kilograms of a supplement is added to each tonne of concentrate, many people find it easier to buy a proprietary mineral mix, which usually supplies additional essential minerals. Manufacturers can usually supply a special mix to meet a particular need.

QUESTIONS

1. Name three major elements found in the mineral matter, and three trace elements.
2. In what two forms are minerals found in the plant or animal?
3. Why are minerals vitally important to living things?
4. How does a plant obtain its mineral supply? Which parts of the plant contain most (a) calcium and (b) phosphorus?
5. Which type of fodder supplies the most minerals? Which type of fodder has the lowest mineral content?
6. In what part of the animal are most of the minerals found?
7. Give an example of a mineral being used in the digestion process.
8. How is the concentration of minerals in the blood controlled by the body?
9. What minerals are deficient in animals suffering from: (a) milk fever; (b) anaemia?
10. What general conditions lead to mineral deficiencies in stock?
11. Which are the major elements supplied by mineral supplements?
12. Name three common substances that are used to supply these elements.

PRACTICAL WORK

1. If you keep dairy cattle or pigs, find out what mineral supplement is being added to the rations and what elements it contains.
2. If you are on an upland farm, are your stock liable to suffer from any deficiency disease? What precautions are taken to prevent this?
3. If you breed pigs, what precautions are taken to prevent anaemia? State the substance used, dose, and time of dosing.
4. What does it cost to add a proprietary mineral supplement to a tonne of concentrate mixture?

CHAPTER 7

The constituents of the dry matter: vitamins

IN COMPARISON with the other constituents the vitamin content of the animal or its food is extremely small. Nevertheless, these substances are just as important as the major constituents. Animals fed on chemically pure water, carbohydrates, and so on, soon suffer from malnutrition and eventually die. This was not realised until the beginning of this century when it was found that other substances had to be present although only in small quantities. These essential materials have since been isolated and analysed, and are known as the **vitamins**. Some are relatively simple compounds that are now possible to manufacture synthetically.

Vitamins are **organic catalysts**, i.e. they speed up the chemical reactions that go on within living organisms. In other words they stimulate the body's function or metabolism. Some vitamins have a general effect while others control a specific reaction or activity. The chemist divides vitamins into two groups—those that readily dissolve in fats and those that are water-soluble. One difference between the two groups is that the body tends to be able to build up a reserve of the fat-soluble vitamins but it is not able to store the second group in this way. Discoveries of new vitamins have been made in recent years, and some which were thought to be single substances have now been shown to consist of several.

Brief notes on some of the main vitamins are given below.

1. Fat-soluble vitamins

(a) Vitamin A is a general growth-promoting vitamin which also helps the animal to resist disease. It is not present in the plant as

34

such, but the animal is able to build it up from the yellow pigment **carotene**, found in the green parts of plants. Fish-liver oils (cod-liver oil), egg yolk, butterfats, and other animal products also contain a good supply.

(b) Vitamin D_3 is important as it prevents the development of rickets in young animals. It is associated with the uptake of calcium and calcium metabolism in general, and is necessary for **correct bone formation** in all stock, laying birds, and lactating animals in particular. This vitamin is similar to vitamin A in that it occurs in cod-liver oil but not in living plants. As sunlight falls on drying grass, vitamin D is developed and it is thus present in well-made hay. Animals exposed to **direct sunlight** are able to produce vitamin D in their skin.

(c) Vitamin E. A lack of vitamin E may cause **sterility**, and so an adequate supply is important for breeding stock. It is widely distributed in plants and feeds and an animal on a varied diet should not suffer from a deficiency. **Wheat-germ oil** is a particularly rich source and it is sometimes recommended in a case of sterility in sows.

2. Water-soluble vitamins

(a) Vitamin B_1 (thiamin) is known as the anti-neuritic vitamin. It acts as a general catalyst to metabolism and it is found in millers' offals (weatings, bran) and plant leaves; and in the animal in the liver, heart, kidneys, and egg yolk.

(b) Vitamin B_2 (riboflavin) is also a **growth-promoting** vitamin and it is present in considerable amounts in green plants and cereal and legume grains. In the animal it is found in egg-white and generally throughout the tissues and organs. The bacteria in the rumen can synthesise both these vitamins.

(c) Vitamin C (ascorbic acid) is only important to human beings as it prevents the skin disease scurvy. This scourge was prevalent in the days of sailing ships when the crews went for long periods without fresh vegetables. Vitamin C is present in fresh fruit and green leafy plants. Farm animals are able to synthesise their own supply and it need **not** be considered in rationing.

PRACTICAL CONSIDERATIONS

In practice vitamins are usually fed in the form of vitamin supplements. They are added to the concentrate rations of animals kept under **intensive conditions**, such as pigs, poultry, and milking animals, and other stock being reared indoors. **Winter time** is the danger period when fodder is not as good as it could be, and stock are housed away from direct sunlight. The supplements are added mainly as a general insurance rather than to prevent the deficiency of any one vitamin.

In practical feeding the two main vitamins are **A and D_3**—the first as a general growth promoter and to help the stock resist infection, the second to encourage calcium metabolism and the maintenance of healthy bones. Vitamins A and D_3 are found in **cod-liver oil**, which is widely used as a vitamin supplement. Alternatively, they can be produced synthetically and added to the ration in crystalline form.

Vitamin B_2 (riboflavin) is also a growth stimulant, and is generally recommended for fattening pigs. It can be added as a proprietary supplement.

As with minerals, the cost of adding vitamins to rations is very small, whereas the losses from unthrifty stock suffering from a deficiency disease can be very great. This alone warrants the addition of supplements to the rations. On the other hand, if the basic principles of good husbandry are followed, many troubles should not arise. It is when too much is expected of stock, either by trying to stimulate production too far or by trying to make do with inadequate housing, or rations that are not up to standard, that these deficiencies are likely to show themselves.

QUESTIONS

1. What is the general action of vitamins in the body?
2. From what substance can the animal manufacture vitamin A?
3. Why is sunlight important to stock?
4. What is the purpose of giving wheat-germ oil to a sow?
5. Which vitamins are synthesised in the rumen?
6. Which vitamins are found in cod-liver oil?
7. What classes of stock may need vitamin supplements adding to their rations, and under what conditions?

PRACTICAL WORK

1. If you are feeding a vitamin supplement to any of your stock, find out:
   (a) what vitamins it contains;
   (b) how much is added per tonne of feed;
   (c) how much it costs to add this to the diet.
2. Obtain manufacturers' leaflets for their feed supplements and note which vitamins are recommended.
3. Write out the details of symptoms, treatment, and so on, of any stock you know that have suffered from a vitamin deficiency.

CHAPTER 8

Digestion and the digestive systems of farm animals

DIGESTION is the process by which the complex substances found in the food are broken down into simpler substances. The animal is then able to absorb these through the wall of the gut and use them in its **metabolism**. Some simple compounds such as glucose or common salt require no digestion, and can be absorbed as they are, but the more complex constituents require digestion. Proteins are broken down to amino acids, fats to fatty acids and glycerol, and carbohydrates into simple monosaccharides (see previous chapters).

The breaking-down process is a series of chemical reactions with a number of intermediate products. Each stage of the process is brought about by the action of an **enzyme** which is a special substance produced by the digestive system for this purpose. Thus there is a series of enzymes which break down protein—the **proteolytic enzymes**. Likewise, **lipolytic enzymes** break down fat and **amylitic enzymes** break down carbohydrates.

One carbohydrate that cannot be digested is cellulose, as there is no enzyme to bring this about. Grazing animals consume vast amounts of cellulose and in order to break it down and make use of it, these animals have to rely on bacteria. In the ruminant animal, **bacterial digestion** takes place in the rumen or paunch. Millions of these single-celled organisms attack the cellulose and break it down to simpler compounds, some of which they utilise themselves. The animal is then able to make use of the simplified byproducts and also digests many of the bacteria.

The digestive system of any animal consists of two parts. First there is the **alimentary canal** (or digestive tract), which is a continuous tube running from the mouth to the anus. In parts the tube is dilated to form muscular sacs such as the stomach. In some sections it is small in diameter, in others relatively wide. For the most part it is extensively coiled, but held in place in the body cavity by a membrane known as the **mesenteric membrane**.

Secondly, there are the **accessory organs** of the digestive system. These organs produce the digestive juices containing **enzymes** and pour them into the alimentary canal at various points along its length, gradually completing the digestive process. Obvious examples are the salivary glands and the pancreas. The lips, tongue, and teeth also play their part in preparing the food for digestion and may be considered as accessory organs.

The basic parts of the alimentary canal and its accessory organs are listed below:

The parts of the alimentary canal		Accessory organs
Preparatory section	(a) Mouth	Lips, teeth, tongue
	(b) Pharynx	salivary glands
	(c) Oesophagus (gullet)	
Digestion and absorbtion	(d) Stomach	
	(e) Small intestine	Liver and pancreas
	(f) Large intestine	
Egestion	(g) Rectum and anus	

The **mouth** is a cavity bounded by the upper and lower jaws. The roof of the mouth is formed by the hard palate which is continued backwards in the soft palate. The tongue forms the floor of the mouth and lies within the mandibles (teeth) of the lower jaw. The opening of the mouth is surrounded by the lips. There are three sets of glands producing saliva.

The lips, tongue, and teeth are used by the animal to take in the food, control it while it is in the mouth, and prepare it for digestion by grinding and mixing with the mucous from the salivary glands. The shape of the mouth gives a good indication of the kind of food the animal eats and the means of acquiring it.

The horse has a normal pair of lips but the top one in particular is very sensitive and mobile. This lip is used to "feel" the food and guide it towards the mouth. The incisor teeth are able to crop herbage closely with a scissor like action. The horse is well supplied with molar or grinding teeth with which to masticate the food before swallowing.

The cow and the sheep, which are also **herbivorous** (grazing) animals, have no incisor teeth in the top jaw, just a hard pad—the dental pad. In this case, the tongue is used to guide food into the mouth, the herbage being gripped between the lower incisors and the dental pad, then torn away.

In the pig, the lips and jaws are greatly modified to form a snout which enables the pig to "root" for its food. The snout is very sensitive and in combination with a keen sense of smell it is able to comprehend its food. The lower incisor teeth are almost horizontal and the lower jaw acts as a shovel. The tusks of the pig are well known and these, like the large canine teeth in the dog and other **carnivorous** (flesh-eating) animals, are used to tear the food and prevent it falling from the mouth. The molar teeth are not as well developed in the pig, indicating that this animal is not designed to graze, but to eat a variety of food both vegetable and animal. Such an animal is said to be **omnivorous.**

The jaw movements vary in herbivorous and carnivorous animals. The herbivorous grind their food and the bottom jaw works with a rotary horizontal movement. Carnivores, on the other hand, merely bite their food with a vertical movement of the jaws and often bolt it without much chewing.

There are three pairs of **salivary glands.** One pair lie under the tongue, another pair in the hind part of the cheek at the base of the ear, and the third smaller pair on the inside of the lower jaw. These glands secrete a watery solution containing **mucin** and in some cases a small amount of an enzyme **ptyalin**, which is an amylitic enzyme starting the breakdown of starch into sugars. A small amount of saliva is secreted continuously but when the animal thinks about food and takes food into its mouth, saliva is produced in considerable quantities. Herbivorous animals produce relatively more than carnivores—a horse may produce 50 litres of saliva in 24 hours, a cow 60 litres. In pigs it is only 1–2 litres.

The **pharynx** is a short muscular tube at the back of the mouth and is a common duct for the passage of food and air. This tube is normally closed to the passage of food but allows air to pass from the nostrils and nasal cavity down the trachea (wind pipe). When an animal swallows, it stops breathing, the soft palate rises and blocks the exit from the nasal cavity and at the same time the **epiglottis** rises and blocks the end of the trachea. The way is then open for the food to cross the pharynx and pass down the **oesophagus.**

The **oesophagus** or **gullet** is a muscular tube that runs from the pharynx to the stomach. It passes down the neck behind the trachea, above the heart, between the lungs, and finally through the diaphragm into the **abdominal** cavity, where it joins the stomach. The length of the oesophagus varies with the size of the animal. In an adult horse it can be up to 1.5 m long. Food is forced along the oesophagus by a wave-like motion of contractions and relaxation of the muscles in the wall of the tube.

The stomach in the horse and the pig is a simple V-shaped muscular sac. At each end are strong circular muscles known as **sphincter muscles**. These control the flow of food coming into the stomach from the oesophagus and again when it passes out of the other end into the small intestine.

The **ruminants** (cattle and sheep) have a compound stomach consisting of four parts. These are the **rumen, reticulum,** and **omasum** through which the food passes before reaching the **abomasum,** which corresponds to the true stomach in other animals.

The stomach acts as a reservoir for food and is never quite empty. Due to contractions of its muscular wall, it has a churning action on the food which helps to break it down in finer particles. In a finer state, the food is more easily attacked by the digestive enzymes. The digestive process actually starts in the stomach. Two-thirds of the stomach wall carries glands that secrete **gastric juice**, which contains mucin together with considerable amounts of **hydrochloric acid** and the enzyme **pepsin**. Pepsin requires an acid medium in which to work; thus in the presence of hydrochloric acid the first stage of protein breakdown begins. Another enzyme, **rennin**, is also present in gastric juice, and this has the power to clot milk. This is important in suckling animals as it prevents the milk from passing through the stomach too quickly giving the pepsin time to act on the milk protein.

From time to time the sphincter muscle guarding the outlet of the stomach relaxes, and a small amount of food, which is now in semi-liquid state, is squeezed into the **duodenum**, which is the first loop of the **small intestine**. From this point onwards the digestive process is functioning to the full. Ducts enter the duodenum from the **liver** and the **pancreas**, bringing digestive juices from these organs to join those secreted by the glands in the walls of the small intestine itself. In this section of the alimentary tract there is the whole range of proteolytic, amylitic, and lipolytic enzymes.

The small intestine is a coiled tube varying in length from 15 m in the pig to 40 m in the cow, with a variation in width from 2.5 cm in the pig, to 10 cm in the horse. Again, this tube has layers of longitudinal and circular muscles which enable it to contract and expand and force the food along. The inner lining is greatly folded and drawn up into finger-like processes called **villi**, which contain **lymph vessels** and many tiny **blood capillaries**. It is through the villi that the simplified food con-stituents are absorbed. A high proportion of the fatty acids and glycerol find their way into the lymph ducts where they are immediately reconsti-tuted into fat. The amino acids and glucose are absorbed by the blood capillaries and pass into the blood-stream.

The last part of the alimentary canal is the **large intestine**, which is shorter in length than the small intestine but much greater in diameter. There are several well-defined parts—the main tube or **colon** terminates in a short tube, the **rectum**, in which waste material, the **faeces**, collects before being expelled through the **anus**. At the point where the small intestine joins the colon, there is a branch known as the **caecum** or **blind gut**, which is concerned with bacterial digestion and is important in herbivorous animals, especially the horse. The caecum is comparatively reduced in ruminants and only poorly developed in the pig. In man, the caecum is only vestigial (i.e. it has become redundant) and is well known as the **appendix.**

During the process of digestion, the body pours a large quantity of liquid into the alimentary tract. If this was lost from the body, the animal would rapidly become dehydrated. It is the function of the large intestine to reabsorb this moisture, and under normal conditions the faeces finally expelled contain only about 70–80% moisture. In times of ill-health this function may be impaired and scouring may occur with considerable loss of moisture from the body.

Apart from water, the faeces contain undigested food, the residues from various digestive secretions, a certain quantity of dead cells from the wall of the gut, and a large quantity of dead bacteria. The faeces of herbivorous animals contain a far larger amount of undigested food than those of the carnivores, the volume of faeces produced by herbivores being proportionately greater.

COMPARISON OF THE DIGESTIVE SYSTEMS IN FARM ANIMALS AND THE PRACTICAL IMPLICATIONS (Fig. 1)

Non-ruminants

Pigs

Of all farm animals, the pig has the simplest digestive system. It has a simple stomach (*monogastic*) followed by a small and large intestine, comparable to that found in man. The caecum is poorly developed. Obviously, the pig is not equipped for digesting fibre as there is no place in its alimentary tract where bacterial digestion can take place. This fact must be borne in mind when rations are being formulated for pigs, and these should not contain more than 5% crude fibre for adult pigs and less for younger pigs. Pigs must not be expected to obtain very much of their nutritional requirements from grazing, and what grass they consume should be young and low in fibre.

Horses

The horse, on the other hand, is better equipped for dealing with fibre, having a well-developed **caecum** into which the undigested fibrous part of the food is passed. Mucin is produced by the caecum but no enzymes. The soaked macerated food is readily attacked by bacteria, the cellulose is broken down, and the byproducts, mainly acetic and butyric acids, are absorbed. Carbon dioxide and methane gas are also produced, and these are eliminated via the anus. The caecum and colon of the horse, where the bacterial fermentation takes place, have a much smaller capacity than the rumen of the cow or sheep, and therefore, the horse is relatively less efficient at dealing with fibrous feeds. For this reason the fibrous feed fed to a horse must be **hay of the best quality**, with a minimum amount of crude fibre and a maximum amount of protein and other food constituents.

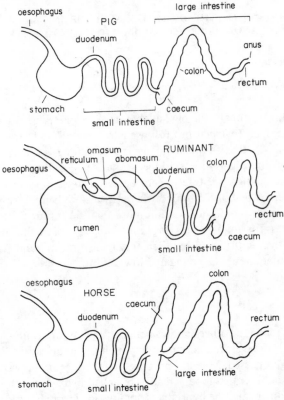

FIG. 1. Diagrammatic representation of the alimentary canal of the pig, ruminant (cow and sheep), and horse.

Ruminants

Cattle and Sheep

The cow and sheep have a compound stomach divided into four sections, which are closely linked together and work as a single unit.

The entire stomach system in adult cattle has a capacity of 150–200 litres and occupies three-quarters of the abdominal cavity. The **rumen** is by far the largest single section, as the following percentages show:

Rumen   80%
Reticulum   5%
Omasum   6–8%
Abomasum   7–8%

The oesophagus enters at the joining between the **rumen** and **reticulum**. The wall of the rumen is modified at this point to form a groove (rather like a gutter) that leads directly into the **omasum**, and is known as the **oesophageal groove**. It is not used in adult animals, but in very young calves milk passes along this groove directly into the omasum and from there into the **abomasum** (true stomach), bypassing the ruman and reticulum. The process is particularly marked in **suckling** calves, but when calves are bucket fed the system does not work so efficiently. By this method, milk, which contains no cellulose, passes straight to the point where normal digestion can commence. The milk-fed calf has no immediate use for the rumen and in the new-born animal it is by no means fully developed. Only when the calf starts to take solid food does the rumen development commence, and so the first food should be of good quality with a relatively low amount of fibre. As time goes on and the rumen grows in capacity, the animal is able to deal with greater quantities of coarser fodder. It is important, therefore, to get the calf on to solid food at an early age. When this does not happen rumen development is slow and it is proportionately longer before the animal can utilise the cheaper farm foods. The early weaning system of calf rearing encourages early development of the rumen.

Cows and sheep are well supplied with molar (grinding) teeth which enable them to chew their food either before swallowing or on regurgitating later. The drier the fodder (hay, for example, as against silage), the longer it is chewed. The food is then swallowed and passes into the upper part of the rumen. Heavy food slowly settles to the lower part. The large amount of liquid in the rumen is mainly saliva secreted in the mouth and is essential if the rumen is to function properly.

The rumen and reticulum working together mix the food thoroughly, bringing it to a semi-liquid state. There are definite waves of contraction which force the food from the lower part of the rumen over into the reticulum, and from there back into the upper part of the rumen to complete the cycle.

There is a certain amount of regurgitation or **cudding**, during this time especially when the fodder is long and coarse. Food is drawn into the oesophagus which forces it back into the mouth. Here it is re-chewed slowly and deliberately, then re-swallowed. A cow spends from 7 to 10 hours a day cudding.

While food is in the rumen and reticulum, bacterial digestion is taking place. Millions of bacteria and other minute creatures attack the cellulose fraction of the animal's food, breaking it down to obtain energy for their own growth and reproduction. The warm, moist conditions are ideal for bacterial fermentation and the food remains in the rumen up to 60 hours. The immediate byproducts are the **fatty acids, acetic** and **butyric**, which are absorbed through the rumen wall and pass in the blood-stream to the liver. Gases are also produced in large quantities— mainly **carbon dioxide** and **methane**, and these are eliminated by belching. If, for any reason, an animal is unable to belch, then the gas pressure increases and the animal becomes **blown**. It is probable that a certain amount of roughage in the food helps the animal to belch. For this reason stockmen may feed some long fodder before turning cattle out to lush young grazing in the spring, when bloat is a possibility.

After a time when the food reaches a certain degree of liquidity the reticulum contracts in the opposite direction and forces the food, together with the bacteria it contains, into the omasum, or third stomach. The omasum is an oval, flattened organ, and internally it consists of about a hundred longitudinal folds, which are roughened by many **small papillae** (projections). Liquid foods fall to the floor of the omasum and pass along quickly into the true stomach. Solid or semi-solid foods become caught between the folds and their papillae, water is squeezed, and some grinding takes place. Ultimately the finely divided food particles reach the fourth stomach where the normal enzymatic digestion commences.

When a ruminant animal swallows some foreign body such as nails, glass, or sand, these tend to accumulate in the reticulum while cinders and stones become lodged in the omasum. The reticulum lies alongside the diaphragm, and on the other side of this is the heart and lungs. If a sharp object such as a nail or piece of wire lodges in the reticulum, it may easily penetrate these organs, the result being fatal.

To summarise, there are two forms of digestion in the farm animals—**enzymatic** and **bacterial**. In the pig digestion is purely enzymatic and it is unable therefore to utilise fibrous foods. With the herbivorous animals digestion is both enzymatic and bacterial, but in the case of the horse, bacterial digestion occurs in the caecum and large intestine, whereas in the cow and sheep it occurs mainly in the rumen. Consequently, cattle and sheep are more efficient than the horse in being able to utilise coarse fodders with a higher fibre content. Table 1 shows the various capacities of each section of the gut in the different farm animals, which underlines these points:

TABLE 1.

Percentage capacity of the sections of the intestines

	Stomach (%)	Small intestine (%)	Large intestine (%)
Horse	9	30	61
Cow	70	19	11
Sheep	67	21	12
Pig	29	34	37

QUESTIONS

1. Into what simple substances are the following food constituents broken down during digestion: (a) protein; (b) fats; (c) carbohydrates?
2. Name the special substances that bring about digestion.
3. How is cellulose digested?
4. Give another name for the digestive tract.
5. Name the main parts of the digestive tract.
6. Name three accessory organs of the digestive system.
7. What names are given to (a) grazing animals, (b) flesh-eating animals?
8. Approximately how much saliva does a cow produce in 24 hours? How does this compare with the pig?
9. What is the function of the oesophagus?
10. Name the four compartments of the ruminant stomach.

11. Besides the enzymes pepsin and rennin, another substance is produced by the stomach—what is it?
12. What are villi and where are they found?
13. What is the function of the caecum and where is it found? In which class of stock is the caecum most important?
14. What is the main function of the large intestine—apart from the caecum?
15. Draw diagrammatic representation of the digestive system of the pig, horse, and ruminant, and label the parts.
16. Why is it the pig cannot digest fibre?
17. Why must fibrous feeds fed to horses be of the highest quality?
18. Of the four parts of the stomach of a ruminant, which is the largest and what is the capacity (in litres) of the stomach system of adult cattle?
19. What is the oesophageal groove and what is its purpose?
20. How long does a cow spend cudding each day?
21. How long does food remain in the rumen and reticulum—what happens to the cellulose during this period?
22. What gases are produced which may cause an animal to be blown?
23. What action has the third stomach (omasum) on the food?
24. Why might it be fatal if an animal swallows a piece of wire?

PRACTICAL WORK

1. Inspect the alimentary canal of pigs and sheep at a slaughter house and note the relative proportion of the various parts.
2. Carefully inspect the compound stomach of a sheep and note the relative size and position of each part.
3. Calculate the percentage of fibre in a typical pig ration and a typical ration for a dairy cow.

CHAPTER 9

Feeds commonly used in livestock feeding

IT IS one thing to know the protein and energy value of a feed, but it is another thing to know its real value as a fodder. Making up a ration, working purely from analysis tables, is rather like selecting breeding stock using only their records. A stock feeder must know much more about the farm feeds than appears in the tables. A ration blended according to the feeding standards is useless, for example, if it is too unpalatable to be eaten. Many feeds have particular characteristics, which make them especially useful to certain classes of stock or at certain times in the year. Linseed cake has a value that cannot be accounted for merely by looking at its protein content, and this is reflected in the price per tonne. Mangolds are greatly prized in the early spring for keeping the milk in suckling ewes, yet an analysis shows only a lot of water and a small amount of carbohydrate. Equally well, many feeds fed at the wrong time can have detrimental effects. A competent stock feeder must have an intimate knowledge of all these facts, and it is at this point where the skill and craft in stockfeeding meet with the scientific approach to the problem.

The following are brief notes on the major characteristics of feeds in common use.

Feeds can be conveniently classified as follows:

(1) Coarse fodders (roughages): bulky feeds with a high crude fibre content and low moisture content.
(2) Succulent feeds (roots): bulky feeds with a high moisture content.
(3) Green fodders: bulky feeds fairly high in moisture content and fibre content.

(4) Concentrate feeds: low moisture content and a high feeding value.

I. COARSE FODDERS

These are cheap, bulky feeds produced on the farm, and consist of the various hays and straws. Some fibrous food of this sort is essential for ruminants, and forms the basis of many maintenance rations for dairy and beef cattle.

1. Hay

Hay is a very variable product and its quality depends upon several factors.

(i) *Quality of the sward.* There are broadly three types of sward from which grass is cut for hay—permanent meadow (meadow hay), temporary grass or ley (seeds hay), and special legume mixture, such as a lucerne-grass mixture (legume hay). The quality of meadow hay varies with the grasses and clovers present in the sward, and can range from a very palatable leafy hay to a coarse, fibrous "herby" mixture. Meadow hay at its best has excellent feeding value, and may not be as coarse as some "seeds" hay, and for this reason is often preferred to seeds hay for feeding to calves and young stock.

Seeds hay and legume hay generally have a higher feeding value, which is due mainly to the presence of the clover and other legumes. The mineral and vitamin content is usually greater than that of meadow hay, the only exception being particularly herby meadow hay which may have had an exceptionally high mineral content.

(ii) *The maturity of the grass when cut.* The stage at which the grass is cut is more important by far than the actual composition of the sward. As grass matures and runs to seed it loses much of its leafyness and becomes stemmy. With maturity the protein content decreases and the fibre content increases. What is equally unfortunate is that the fibre also changes from digestible cellulose to indigestible lignin, and the overall digestibility and thereby the feeding value falls dramatically. Grass is best cut for hay at the early flowering stage, when the maximum bulk of high quality material is obtained.

(iii) *Success in making*. Hay that is made quickly under ideal conditions generally has the highest feeding value, although very good hay can be made by allowing the material to "make" slowly on tripods. Many of the nutrients are lost if the grass stays green after cutting or if it lies for several days being washed out by the rain and bleached by the sun. Under such conditions, a black dusty material is produced with more stem than leaf, and the feeding value is reduced to a minimum.

Good hay should have light green colour with plenty of leaf and a sweet (non-musty) smell.

Hay is fed to all classes of cattle, sheep, and horses. Very few farmers have completely replaced their hay by silage, and for many it still remains the basis of the winter feeding programme in dairy herds. The quality of the hay from one season to the next can have a great effect on the overall milk yield, and the reliance that has to be placed on concentrates to maintain yield.

Hay fed to horses must be of the highest quality.

2. Straw

Oat and barley straw are the straws that are used for feeding. Wheat straw is too fibrous and unpalatable to be used in any quantity. Oat and barley straw both have an energy value approaching that of poor quality hay, but the amount of protein is very small. Straw from corn that is undersown may have a greater feeding value if the undersown seeds have taken well and grown up into the straw, as so often happens in a wet season. The type of straw used depends largely on the availability. Barley straw is normally available in most parts but oat straw may also be available in the north and west.

In general, straw is more filling than fattening, and forms a useful part of the ration for store cattle and dry cows, and can be used to replace some or all of the hay for fattening bullocks. It should not be used for milking cows except for the lowest yielders.

Where peas and beans are grown, the haulm may be used for store cattle, but it is very fibrous and not very palatable, although higher in protein content.

II. SUCCULENT FEEDS OR ROOTS

The root crops, turnip, swedes, mangolds, and fodder beet are also home-grown cheap feeds, but with a low, dry matter content, and therefore a low feeding value. They are used as the basis of diets for cattle and sheep and tend to have a **laxative** effect.

It is common practice to slice roots to make feeding easier, particularly where large quantities are being consumed. Care should be taken when roots are frozen or badly soiled. Under these conditions serious digestive disturbances may occur. Large quantities of roots should not be fed to ewes for the last few weeks before lambing.

1. Turnips

White-fleshed varieties contain only 7–8% dry matter and they are grown chiefly in northern districts for sheep folding. The yellow turnip has a higher dry matter and is similar to the swede in having a higher feeding value and better keeping quality. Turnips cause a milk taint so they are not normally used for milking cows.

2. Swedes

The dry matter is higher in swedes than in turnips (10–13%), and they are richer in protein. It is usual to lift them in the north, but they are fairly winter hardy. They can be fed to all classes of stock, and cattle may consume up to 50 kg if the roots are pulped. The milk may be tainted when swedes are fed to dairy cows. They are ideal for sheep folding.

3. Mangolds

Mangolds are a safe palatable food for all classes of stock, although time must be allowed for them to mature after lifting before being fed. Variation occurs in the dry-matter content between varieties, the yellow globe variety being the poorest. Mangolds do not taint the milk and they can be safely fed to dairy cows, 20 kg per day being the optimum amount.

4. Fodder Beet

Fodder beet is a suitable crop for the drier areas. There are several strains, and although the dry matter is generally high it is variable (17–23%). When cereals are in short supply or are expensive, fodder beet makes a useful supplementary feed for pigs, and it is particularly useful as a change of diet for breeding stock when fresh grazing is not available. Fodder beet is not usually fed to cattle but when it is, a variety with a low dry-matter content should be used.

5. Potatoes

Potatoes have a high dry-matter content, which is mainly carbohydrate (starch) with little protein or minerals. Raw potatoes are not suitable for any class of young stock or for pigs. A toxic substance **solanin** could be present if the tubers have started to turn green through exposure to sunlight. (Boiling destroys the solanin.) They may, however, be given to dairy cows, and fattening cattle or sheep. Quantities should be restricted at first (2 kg for cattle, 0.5 kg for sheep) due to their laxative action.

Maximum quantities: Sheep, 2.5 kg

Dairy cattle, 9–13 kg

Bullocks, 18 kg

Boiled or steamed potatoes can be profitably fed to pigs and poultry providing a regular adequate supply can be obtained cheaply.

Four kilograms of boiled potatoes can replace 1 kg barley meal, and they can be fed to pigs from 16 weeks to bacon weight, beginning at 1.5 kg per day to a maximum of 5 kg per day.

III. GREEN FODDERS

In this group are the true green crops, kale, cabbage, and rape, and also grass, either fresh or conserved as silage. Sugar-beet tops and swede tops must also be classified in this section.

Kales, rape, cabbages—These crops are very valuable in providing fresh green fodder during the autumn and winter months. They all have a high feeding value and they are rich in minerals and carotene (vitamin A). The high water content makes them a laxative food, and for all these reasons they help to keep stock in a healthy condition during the autumn and early part of the winter when fresh grazing would not be otherwise available.

1. Kales

There are three varieties of kale: marrow stem, thousand head, and hungry gap.

Marrow stem kale is the most popular, giving a heavy yield of fresh green fodder during the autumn months. It is not winter hardy, and heavy frosts after Christmas can drastically reduce the yield, especially in the northern counties. Under good management, a yield of 70 tonnes per hectare can be obtained. Where the land is not too heavy and well drained, kale can be strip-grazed by dairy cows with very little waste. Alternatively, it may be desirable to cut and cart it to the stock to avoid puddling the ground when conditions are heavy and wet.

Kale is usually reserved for dairy cattle and large quantities per day may be consumed. A good supply of kale (25–35 kg) and a forkful of hay (3–4 kg) will supply enough nutrients for maintenance and five litres of milk for an average-sized cow. All classes of stock benefit from this fresh green food in the Autumn months.

Thousand-head kale is a variety that will withstand winter conditions. It has a growth similar to brussel sprouts and as the spring approaches the "sprouts" develop into leafy shoots and the yield increases. Sheep are often folded on this crop but it is equally suitable for cattle. In the hardier districts it is probably preferable to house cattle in the depth of winter, and other fodders such as silage can be fed more economically. There is also some evidence that prolonged overfeeding of kale can lead to anaemia in dairy cows.

Hungry gap kale is the latest variety suitable for late winter to early spring use—similar characteristics to Thousand-head.

2. Rape

Rape is popular in many areas as a nurse crop when re-seeding, especially for an August re-seeding. It is very palatable and is usually folded off with cattle or sheep. Rape is ideal for "flushing" ewes or finishing lambs, and it is safe feed for all stock.

3. Cabbages

There are early and late varieties of cabbage and these may be ready for feeding from August, through to the early spring. They are used extensively in the late summer for keeping cattle and sheep in show condition, and later as a general green fodder for all classes of stock. One drawback to cabbages is that they cannot be stored, and if they are not eaten off when they reach maturity they are liable to rot. There is also a chance the milk may be tainted due to the intense smell that may develop when they are fed under cover.

4. Sugar-beet tops

In the arable areas sugar-beet tops are a very valuable fodder from October to Christmas. Later in the winter, hard frost may reduce the yield considerably. The tops, which include part of the crown of the beet, are a laxative palatable food suitable for most classes of stock, but they should be allowed to wilt before use. Good quality tops have a feed value similar to kale. Sheep, bullocks, and store cattle are often folded on tops, which may also be carted off and fed to stock on grass. Care should be taken in hard frosty conditions and where the tops have been soiled.

5. Swede tops

Swedes are usually lifted and stored before feeding, and the tops are then available for feeding during the lifting period. They have a similar feed value to the root of the swede but they are not quite as good as kale. It is common practice to cart them off and feed at grass.

6. Grass

Grazing in the form of grass, grass and legume (clover) mixture, or pure legume crops, forms the major part of the diet of ruminants and other stock during the summer months. The grazing season has been lengthened in recent years by the use of early and late strains and it is possible to graze stock from 6 to 10 months of the year depending on the area and soil conditions.

All grasses are most nutritious and palatable when in full leaf from 10 cm to 25 cm high. As grass matures and goes to seed, the proportion of fibre increases and the proportion of protein decreases correspondingly. The better strains stay leafy and palatable longer than the poorer strains, and this accounts for much of the difference in value between the poorer permanent pastures and the better pastures and leys.

A percentage of clover in a sward always increases the feeding value providing it does not become dominant, when total yield is reduced. A good sward contains the right balance of the leafy palatable species of grass and the clover. Care must be taken to watch for "bloat" when stock are turned on to lush pasture containing a high proportion of clover. Grass always has a laxative effect, but lush spring grazing, low in fibre, may cause excessive scouring in cattle. It is inadvisable to turn out cattle that have empty stomachs on to this type of grazing.

It is also under such conditions, which involve a sudden change of diet, that the mineral metabolism of an animal can be upset, and **grass staggers** (hypomagnesaemia) can occur in cattle. To prevent this disorder additional magnesium is fed in the form of calcined magnesite before and during the turning out period. The calcium in the young grass is then balanced by the extra magnesium and allows the animal time to settle to its new diet.

Autumn grass (foggage) never appears to have the same feed value as spring grass. This is probably due to the wet dewy conditions in the autumn, the stock having to consume large quantities of water along with the grass.

7. Silage

The feeding value of silage depends on the quality of the material used, the stage of growth at which it is cut, and how well it is made. First-class silage can only be made from first-class leys, cut when the grasses are just showing their flower heads. Wilting in the field will help to reduce the water content and the aim should be for a final dry matter of around 25%. The fermentation process should be controlled by the exclusion of air by rolling and by sheeting up each night. Additives may also improve the process particularly when the weather is wet. The final sealing of the clamp is also important.

Even if a good dry matter has been achieved, the content of the dry matter in terms of protein and energy value can vary greatly, and before feeding commences it is as well to have a sample analysed. At the end of the day, however, the only true analysis is how well the stock respond when it is fed to them.

The ways of utilising and feeding silage are almost as varied as the types and qualities of silages. On many dairy farms silage has now become the main bulk fodder for feeding through the winter, the cows consuming up to 50 kg per day. With many other farmers, silage has replaced the roots for feeding to dairy cows or bullocks and sheep, from Christmas onwards. Feeding techniques vary from carting out to stock at grass, to self-feeding from the face of the silage in the clamp, but when the clamp is opened and the silage exposed to the air care must be taken to see that secondary fermentation does not take place.

Tower silage

A tower offers an opportunity to make silage of the highest feeding value. The dry-matter content is higher and the target is about 40%. In a tower the surface area is small in relation to the volume of silage and the pressure of the material excludes most of the air. The carbon dioxide which is produced is unable to disperse even in an unsealed tower, and the material is preserved in an atmosphere of carbon dioxide. If the weather conditions are good to give rapid wilting, the loss of nutrients may be as low as 5%.

On the feeding side the higher dry-matter tower silage should have a higher feeding value, weight for weight, and must be rationed accordingly. In calculating what silage is available for feeding a volume measure is often used. It must be remembered, however, that because of settling, it is virtually impossible to have a tower completely full of settled silage, and some towers may be found to be only two-thirds to three-quarters full when opened up. If top unloading machinery is used, then the usage rate (in terms of centimetres per day) must be enough to prevent mould growing down into the silage. From 8 to 15 cm per day is the recommended rate.

With both tower silage and good clamp silage the protein content can often be quite high—high enough in fact, only to require barley to be added to give a properly balanced diet, even for high-yielding cows. The need to purchase expensive high protein feeds is then reduced or even eliminated.

IV. CONCENTRATES

The concentrates are those feeds which have a high dry-matter content and a very high feeding value. They are expensive and great care must be taken to feed them without waste. The high feeding value may be due to:

(a) a high protein content (protein concentrates), or
(b) a high energy content (carbohydrate or high energy concentrates).

A protein concentrate may also have a high energy content. Dried separated milk powder has an energy content as high as barley, and white fish meal almost as much as oats. The milk powder and the fish meal are classified as high protein feeds, whereas barley and oats are high energy feeds. In general the grouping of concentrate feeds is done mainly on a protein content, and it just so happens that the cereals, including maize, which are lowest in protein, have a relatively high energy content. Furthermore, it would be uneconomic to purchase high protein feeds, which are the most expensive, if all that was required was a high energy feed.

Some concentrates can be grown on the farm—homegrown. By far the most important is barley, which with the other cereals is a high energy feed. Peas and beans are also grown in some areas and are classified as a medium protein feed. Other concentrates come from crops grown overseas, among them maize and locust bean, but probably the greatest number are byproducts from other industries. Wheat feeds (bran and weatings) are a byproduct of flour milling, and the various oil cakes and meals are the materials left after oil has been extracted from oil-bearing seeds; and there are many others.

Most concentrates are derived from plants, but a number of the most valuable are of animal origin. Fish meal and meat meal are two important examples and an essential source of **animal protein** of **high biological value.**

High protein compounds

For many practical farmers the main source of protein concentrate is a manufactured protein compound. Feeds such as "grain balancer cake" or other aptly named high protein supplements are designed to make home mixing of diets a simple matter. In addition to the protein content minerals and vitamins are usually included in the compound, sometimes known as *protein–mineral–vitamin* (PMV) supplement, and when mixed at recommended inclusion rates with home-grown cereals, provides a properly balanced ration for a particular class of stock. The actual energy and protein value of these compounds is not always widely published and, of course, can be varied from firm to firm. One or two have been included in the list of concentrates at the end of this chapter but the actual analysis as shown is only approximate.

The following is a brief description of each feed grouped according to protein content and origin, with the analysis stated in terms of grams of **digestible crude protein** per kilogram of dry matter (g/kg DM) and energy in terms of megajoules of **metabolisable energy** per kilogram of dry matter (ME/kg DM). An explanation of these terms is given in Chapter 7. Both the DCP and ME for the feeds described are for ruminant stock, and they are summarised at the end of the chapter. A summary of the feeds used in pig feeding is given at the end of Chapter 14.

Group I. High protein feeds of animal origin

(i) WHITE FISH MEAL. DCP 631. ME 11.1

White fish meal is made from genuine "white" fish (e.g. cod, haddock, plaice), which does not contain much oil or strong flavouring. There is a guarantee that the product does not contain more than 6% oil or 4% salt. Other poorer quality fish meals have no such guarantee and are more likely to produce taints in milk or bacon. Besides being a valuable source of animal protein, white fish meal contains a high proportion of minerals. The ground bones and salt present supply the three major minerals phosphorus, calcium, and chlorine.

These features make white fish meal an essential food for all young growing stock. It has an unrivalled place in pig rations where up to 10 or 15% may be fed—although this amount should be reduced as bacon weight is approached to avoid any chance of a taint in the carcass. Fish meal is usually considered to be too expensive for inclusion in the rations of adult cattle and it is not normally used for milking cows. Some authorities are now advocating its use, however, and it may become a regular constituent of milk production rations in future.

Fish meal has a light brown colour with small pieces of paler coloured bone, and it possesses a characteristic smell.

(ii) MEAT AND BONE MEAL. DCP 465. ME 9.7

There are several byproducts from slaughter houses used in feeding, but meat and bone meal and feeding meat meal are the two most important. Both are high protein feeds with a high mineral content and may be fed on similar lines to white fish meal. The high fat content is often considered to be a detrimental characteristic of these feeds, but the "feeding" products can be obtained with a guarantee that they do not contain more than 4% fat. Although they may be used in place of white fish meal they are generally thought to be of lower feeding value. In terms of protein equivalent this appears to be true, although feeding meat meal (low fat) is only slightly inferior in this respect. The final choice is largely governed by the relative price and the regularity of supply of a consistent sample.

(iii) DRIED SEPARATED MILK. DCP 350. ME 14.1

After the removal of the butter fat, skimmed or separated milk still contains the milk protein and minerals. When dried, a valuable highly palatable and easily digested feed is produced. It is a first-class feed for young pigs, calves, and poultry, and up to 5 or 10% may be included in the rations. Once again, cost limits an extensive use of this feed.

Group II. High protein feeds derived from plants

Oil cakes and meals

Oil cakes are manufactured from seeds rich in oil. There are two main ways of removing the oil; by the use of an expeller screw press (expeller cakes) or by means of an organic solvent (extracted cakes). The extraction process is the most efficient leaving only 1 or 2% oil. As a result the final byproducts differ both in analysis and physical appearance. The removal of the extra oil reduces the energy feeding value of the extracted product, but the protein value is proportionately higher. Because of the reduced oil content, the extraction product tends to break down into a meal which renders it less palatable. The addition of molasses may overcome this.

Before the oil is removed it is necessary with some seeds to remove the outer coat or husk. Such seeds are then **"decorticated"** and the final product is a **"decorticated cake"**. When the husks are not removed an **"undecorticated cake"** is produced. Decorticated cakes, without the fibrous husks, naturally have a lower fibre content but a proportionately higher protein content and greater feeding value. It may be an advantage to feed an undecorticated cake under certain circumstances, particularly when the rest of the ration is low in fibre, and there is a tendency for cattle to scour.

(i) SOYA BEAN MEAL AND CAKE. MEAL: DCP 453. ME 12.3
CAKE: DCP 454. ME 13.3

Soya bean meal or cake is the residue after oil has been removed from soya beans. It is a popular feed suitable for cattle, sheep, and pigs. The protein content is twice that of field beans grown in this country. The fibre content is low and either cake or meal is slightly laxative. In pig rations soya bean meal is used as a protein supplement to replace fish meal in the final stages of fattening; but a mineral supplement must then be included, as soya bean, though rich in phosphorus is poor in calcium.

From a "texture" point of view soya bean is a heavy meal and many milk producers limit its use to 10–15% of the production ration.

Appearance—pale, yellow, "gritty" meal.

(ii) GROUNDNUT MEAL AND CAKE. MEAL: DCP 491. ME 11.7
CAKE: DCP 449. ME 12.9

This is the residue after oil has been removed from groundnuts and may be either decorticated or undecorticated. The decorticated cake is very palatable and digestible and is perhaps the most popular protein feed for dairy cows. The undecorticated form has a lower feeding value, is less palatable, and is usually fed to store cattle and low producing cows.

Appearance—middle brown meal or flakes with a smell of peanuts.

(iii) DECORTICATED COTTON CAKE
(SEE COTTON CAKE BELOW).

Group III. Medium protein feed

(i) COTTON CAKE. DECORTICATED: DCP 393. ME 12.3
UNDECORTICATED: DCP 203. ME 8.7

There are again two grades of cotton cake—decorticated and undecorticated, and the term "cotton cake" implies the undecorticated form. It is a useful food for counteracting "looseness" in cattle as it has a costive action due to its high fibre content. Although not very palatable it is traditionally the protein supplement for inclusion in rations for beef cattle. In practice it is not fed to sheep to any extent, nor is it suitable for young stock under 6 months. It should not be fed to poultry.

(Decorticated cotton cake has nearly twice the feeding value and is included in Group II above.)

Appearance—green–brown in colour and recognised by the presence of fibres and pieces of black seed hulls. Smell—sharp, peppery.

(ii) LINSEED CAKE. DCP 286. ME 13.4

Linseed cake has been a popular food for a long time, and even today there is more of it sold than any other cake. Because it is very palatable with a slightly laxative effect it is a valuable food for sick animals. It is also well known for putting "bloom" on to animals for show or sale. On the other hand, it has always been priced relatively high, and in terms of protein other cakes such as groundnut are often a better buy. It may be used in reasonable quantities for dairy cows, suckling ewes and lambs, and fattening cattle and sheep.

Appearance—middle brown, usually with a glossy side. Smell— linseed oil, like putty.

(iii) DISTILLERS' GRAINS (DRIED: 90% DM). DCP 214. ME 12.1

Distillers' grains are a similar product to the more common brewers' grains but they have a higher feeding value. They are a byproduct from whisky distilling and contain varying proportions of barley and maize grains. Individual farmers prize them for milk production, buying them in bulk in the autumn and ensiling them for use in the winter. The dried grains are a useful medium protein food for milk production.

(iv) PEA AND BEAN MEAL. PEAS: DCP 225. ME 13.4
BEANS: DCP 209. ME 12.8

Peas and beans are a useful source of home-grown protein. Providing they are allowed to mature before use they are a safe food for all stock and a stimulant for milk production. Should bean meal become damp it heats and is then unpalatable. Such meal causes scouring, but normal bean meal has a costive action due to a low oil content.

(v) PALM KERNEL MEAL AND CAKE. MEAL: DCP 204. ME 12.2
CAKE: DCP 196. ME 12.8

When molasses is added to palm kernel meal palatability is improved and it becomes a most useful feed for cattle over 6 months, and for dairy cows in particular. It is not suitable for pigs under normal conditions.

Appearance—light fawn with brown bits. Smell—not pronounced.

(vi) DRIED GRASS—FIRST QUALITY. DCP 136. ME 10.6

Dried grass is a variable product containing from 6 to 14% DCP. Good quality dried grass must contain at least 13% protein, whereas maintenance quality may contain less than 13% but more than 10% (these figures assume 10% moisture content). Most classes of stock benefit from having a proportion of dried grass in their rations as it contains vitamin A and acts as a tonic during the winter months. Because of the relatively high fibre content, care must be taken when feeding it to non-ruminants.

Maintenance quality dried grass is balanced* for milk production whereas the first-quality product needs to be balanced with a cereal.

Appearance—dark green, finely ground meal.

(vii) WEATINGS. DCP 129. ME 11.9

Weatings is another byproduct of flour milling, and may also be known as middlings, pollards, thirds, or sharps. It contains less of the skin of the wheat and more of the flour than bran. Weatings is considered to be a very safe feed and is fed to all classes of stock. Besides being included in dairy rations as a balanced feed,* it has an important place in most pig rations. The fibre content is a little variable, however, and merchants are required to stipulate the amount present.

Appearance—like fine bran, but heavier and paler. No pronounced smell.

(viii) BRAN. DCP 126. ME 10.1

Bran is the outer skin of wheat left after the flour has been removed in the milling process. It is a light bulky feed and when fed as a bran mash it has a laxative effect. Compared with other concentrates its feeding value is not very high, but it is a popular ingredient in dairy rations. There are two grades—broad bran and fine bran.

Appearance—large fawn-pink flakes, thin and light. Fine bran—smaller flakes. No pronounced smell.

* balanced – having a similar protein to energy ratio as milk.

(ix) BREWERS' GRAINS. DCP 145. ME 10.3

Brewers' grains, which may be dried or fresh, are the residue of malted and unmalted barley used in the brewing process. Where there is a brewery close to the farm it is often economic to buy fresh grains and feed them wet. Either fresh or dried this feed has the reputation for stimulating milk production. A limited proportion of dried grains may be used in the concentrate mix. As grains are rather fibrous they are not usually fed to pigs or poultry.

Group IV. High energy feeds

(i) FLAKED MAIZE. DCP 106. ME 15.0

Flaked maize is produced by passing steam-cooked maize between rollers. The process gives rise to an extremely palatable and easily digested feed. Because of its flaky non-mealy nature it improves the texture of a meal mixture to which it is added. These characteristics make flaked maize eminently important as a feed for young stock, sick animals, and in any other situation when palatability is of prime importance.

Appearance—yellow–orange, like cornflakes.

(ii) MAIZE. DCP 78. ME 14.2

Most of the maize used in this country is imported. In many countries maize, or Indian corn as it is called, is the main carbohydrate feed for fattening cattle, pigs, and sheep. In this country it is not normally the major ingredient of a fattening ration, but it may profitably be used along with oats and barley. Too high a percentage in a pig ration may cause soft fat in the carcass. All maize feeds tend to be deficient in certain amino acids, but this is not a major detriment in a fattening feed.

(iii) WHEAT. DCP 105. ME 14.0

Wheat is normally sold for milling and it is only the inferior grain that is usually retained for feeding. It contains much less fibre than oats and may cause digestive disturbances by forming an indigestible paste in the stomach. A ration should not contain more than 25% therefore, although in recent years pigs on trial have been fed up to 40% coarsely ground wheat without harm. The nutrient value is higher than any other cereal. Whole grain is often fed to poultry.

(iv) LOCUST BEANS. DCP 47. ME 13.8

Locust beans are sold either kibbled or as a meal. This food contains a fair proportion of sugar which makes it very palatable, and it is often fed with less palatable foods. It is useful for dairy and fattening cattle and especially useful for fat lambs.

Appearance—Dark brown shiny outside, light brown inside. Smell—sweet, a little like molasses.

(v) BARLEY. DCP 82. ME 13.0

Barley has the highest energy value of the home-grown cereals. It is suitable for all stock and especially pigs because of its low fibre content; pig-fattening rations may contain up to 70%. Ground barley contains less husk than oats, and the meal appears white and gritty. It is normally rolled (crushed) before feeding to ruminants.

(vi) MAIZE GERM MEAL. DCP 90. ME 13.2

Maize germ meal (sometimes cubed) is the result of grinding maize germs from which the oil may or may not have been removed. Supplies are naturally limited but it is a very popular feed with a number of farmers both for dairy rations and for fattening rations for cattle and sheep. Similar to barley in many ways, it has somewhat better feeding value and slightly less fibre.

(vii) DRIED SUGAR-BEET PULP. DCP 61. ME 12.2

Sugar-beet pulp is the residue left after sugar has been extracted from beet. Molasses is usually added which improves palatability and slightly increases the energy feeding value. A high fibre content makes it unsuitable for pigs, and as it swells with the addition of moisture it should be fed with care to all stock. Disastrous results may follow feeding even a small amount to horses. Where there is a reliable supply, sugar-beet pulp can replace a proportion of the cereal in a ration and it may also be used as a bulky succulent feed if it is first soaked in water.

Appearance—dry tough shredded, brown or fawn. Sweet smell.

(viii) OATS. DCP 84. ME 11.5

Oats are more fibrous than other cereals and they used to have an important place in the rations of cattle, sheep, and horses. Because of the strong husk, oats must always be crushed. If finely ground they may be fed to pigs in limited quantities.

Other compounds

In addition to the high protein feeds mentioned at the beginning of this chapter, livestock feed compounders normally produce two other types of compound feed:

(a) Complete diets designed to meet the total requirements of the animal. Such diets are feasible in pig and poultry feeding where there is no grazing element and the animal can only take one type of feed, i.e. a concentrate low in fibre. Compounds can be purchased, therefore, for all classes of pigs and poultry, which are complete diets, and when fed at the prescribed rates, meet the daily nutritional needs of the stock for both maintenance and production.

(b) Diets balanced to meet production needs only. Compound dairy nuts and beef-fattening nuts are examples of compound feeds that cater for the production of an animal. The theory is that a certain amount of the compound provides the nutrients for a unit of production—say, a litre of milk or a kilogram of liveweight gain. At the same time the maintenance needs have been met from other feeds and, in these examples, it could be from hay or silage.

Such a system has in fact been traditional for many years, and may still have a place for many farmers; but there are difficulties in that the quality of the home-grown feed may be very variable and in no way match up to the maintenance needs of the animal, nor

TABLE 2.
Feed analysis table: concentrate feeds

	Dry matter (DM) (g/kg)	Metabolisable energy (ME) (MJ/kg DM)	Digestible crude protein (DCP) (g/kg DM)
GROUP I			
White fish meal	900	11.1	631
Meat and bone meal	900	9.7	465
Dried separated milk	900	14.1	350
High protein compound*	900	12.0	450
GROUP II			
Soya bean meal	900	12.3	453
Soya bean cake	900	13.3	454
Groundnut meal	900	11.7	491
Groundnut cake	900	12.9	449
Grain balancer compound*	900	12.5	280
GROUP III			
Cotton cake: decorticated	900	12.3	393
Cotton cake: undecorticated	900	8.7	203
Linseed cake	900	13.4	286
Distillers' grains (dried)	900	12.1	214
Pea meal	860	13.4	225
Bean meal	860	12.8	209
Palm kernel meal	900	12.2	204
Palm kernel cake	900	12.8	196
Compound dairy nuts*	880	12.8	130
Dried grass	900	10.6	136
Weatings	880	11.9	129
Bran	880	10.1	126
Brewers' grains (dried)	900	10.3	145
GROUP IV			
Flaked maize	900	15.0	106
Maize	860	14.2	78
Wheat	860	14.0	105
Locust beans	860	13.8	47
Barley	860	13.0	82
Maize germ meal	900	13.2	90
Dried sugar-beet pulp	900	12.2	61
Oats	860	11.5	84

*The analysis of these manufactured feeds should be checked with the compounder.

to the production compound. The modern approach is to determine the total needs of the animal and see what feeds are available and what additional feed needs to be purchased to produce a balanced diet and ration. It can be seen, however, that, depending on many circumstances, a farmer has the choice of purchasing either:

(i)   a complete feed;
(ii) a feed designed to meet the production needs;
(iii) a high protein (including minerals and vitamins) supplement to mix with his own farm feeds; or
(iv) "straight" feeds plus minerals and vitamins from which to mix the diets required.

QUESTIONS AND PRACTICAL WORK

As this chapter is purely descriptive, questions are not considered appropriate. The analysis for the bulky feeds has not been given, and the DM, ME, and DCP could be found for typical examples in the feed analysis tables and inserted in the margin by each feed.

PART II

Rationing Farm Livestock

CHAPTER 10

Terms used in animal nutrition

BEFORE attempting to compile rations for stock it is essential that the main terms which are currently in use are properly understood. Many of these terms have abbreviations, such as DM for the dry-matter content, and it is as well to be familiar with these, for besides being better able to interpret the nutritional jargon, they are also a convenient form of shorthand.

DRY MATTER (DM)

As mentioned in Chapter 1, the feeding value lies in the DM of the food, the remainder being water.

The dry-matter content. The DM is expressed as so many grams (g) per kilogram (kg) of the food, e.g.:

DM of clamp silage is 200, i.e. 200 g DM/kg silage (as a percentage this is 20%).

Tower silage, on the other hand, has a DM of 400, i.e. 400 g DM/kg (or 40%)—twice as much as the clamp silage.

Dry matter is important to the animal as it is used as a measure of its appetite, i.e. the amount it can eat in a day.

A dairy cow's appetite varies according to its milk yield but it can eat some 2–3 kg DM per 100 kg liveweight. This daily allowance is termed the **dry-matter intake**—(DMI); and the ration must be such that all the required nutrients are contained within this quantity of dry matter.

The DMI of a dairy cow weighing 590 kg and producing 20 kg of milk is given as 17.0, i.e. 17.0 kg DMI/day.

FEED—AS FED

Although it is usually the dry matter which is referred to in feed analysis and in compiling rations, it is also necessary to refer to the **whole feed** as actually fed to an animal. When this is so the term **as fed** is used. If it is just the dry matter, then DM is placed after the feed. For example, if a silage has a dry matter content of 25%, then

1 kg silage DM = 4 kg silage as fed.

FEEDS, DIETS, AND RATIONS

It is as well to be consistent in the use of these three terms. A **feed** is the term given to a particular product, such as barley, maize, or silage. A number of individual feeds are selected for feeding to stock and the group becomes a **diet**. Growing pigs may be fed a diet of soya-bean meal, weatings, and barley in a particular combination. The quantity of the diet fed per day is the **ration**.

These are the terms used in this text. "Feedingstuffs" is a term which is still commonly used when a group or list of feeds is being referred to —as in "feedingstuffs manufacturers" or "feedingstuffs tables".

MAINTENANCE AND PRODUCTION

A farm animal uses its food for either maintenance or production (see Chapter 1). When rationing stock we tend to think in terms of these two needs in order to arrive at the animal's overall needs. These are terms that need to be clearly understood.

The maintenance allowance: the nutrients required by an animal to keep it in good bodily health and to maintain its body temperature without gain or loss in weight.

The production allowance: the nutrients required by an animal, in addition to the maintenance allowance, to supply the nutrients needed for some form of production, e.g. growth, fattening, or milking.

For each of these there are two main nutritional considerations:
(a) a supply of energy;
(b) a supply of protein.

ENERGY

The term "energy" includes the actual physical energy an animal needs, the heat to maintain its body temperature, and the nutrients for laying down its own energy reserve. When an animal has met its normal energy/heat requirements, any surplus is converted and stored as fat. When it is stated that a feed has a certain **energy** value, it could equally well be said that it has a certain **heat** value or **fattening** value—the terms are synonymous.

The constituents that give food its energy value are the carbohydrates, the fats, and the protein. While it is readily recognised that the carbohydrates (sugars and starch) and the fats are energy feeds, the contribution that the protein makes is not always appreciated. A carnivorous animal such as a tiger, whose diet is largely protein, can hardly be said to lack energy. Fish meal, which contains no carbohydrate but which is high in protein, has an energy value equal to oats, which is a typical starchy fattening feed. In these examples it is the protein which is giving the feed most of its energy value.

The measurement of energy feeding value

Gross energy (GE): The total energy in a feed is known as the gross energy (GE). When a feed is burned it produces an amount of heat equal to its GE value. Unfortunately, not all the GE is available to the animal as some of the constituents are indigestible and pass through the intestine and appear in the faeces. The GE is of little use, therefore, as a measure of the energy feeding value of a feed.

Digestible energy (DE): The energy contained in the digestible part of the feed is known simply as the digestible energy (DE). Clearly, this is a more accurate measure of the energy feeding value, and it is this figure that is used in the rationing of pigs. Not quite all the DE becomes available to the pig, however, as a small amount is lost in the urine.

Metabolisable energy (ME): In ruminants also there is a loss of DE in the urine, but there is a much greater loss through the methane gas which is produced in the rumen during bacterial digestion. The energy that remains is available to the animal for use in its own metabolism and is known as the metabolisable energy (ME). It is the ME figure that is used in rationing cattle and sheep, particularly lactating animals, and to a lesser extent for growing stock.

Heat increment and net energy (NE): The ME is used by the animal in two ways. It may be used as a source of energy for the metabolic processes including the process of forming new tissues. This part of the ME is converted into work and heat and is eventually lost as body heat. It is known as the "heat increment". The remainder of the ME is incorporated in the new tissues and is known as the net energy (NE). The NE is, in fact, the net product of the entire energy feeding process and may only be a small part of the GE content of the food eaten by the animal.

The energy feeding process is represented diagrammatically below

The megajoule (MJ): The unit for measuring energy is the megajoule (MJ). The energy needs of animals are expressed as so many **megajoules per day (MJ/day)** and the energy value of feeds as **megajoules per kilogram of dry matter (MJ/kg DM)**. For example:

a Jersey cow giving 20 kg milk per day requires

      163 megajoules of metabolisable energy per day

or 163 ME (MJ/day)

or simply 163 ME.

barley has an energy value of

      13.0 megajoules of metabolisable energy per kilogram of dry matter

or 13.0 ME (MJ/kg DM)

or simply 13.0 ME

Comparing the energy values of feeds: When comparing the energy value of two feeds it must be remembered that the energy value given in feed analysis tables and elsewhere relates to the DM of the feed and not to the feed as fed. If the ME value of two feeds, for example, is similar, their value as feeds as fed will depend on how much DM each contains. Samples of hay and silage may have a similar ME, e.g.:

hay   8.4 ME
silage   9.0 ME

but hay has 85% DM whereas good silage has only 25% DM, which means:

$$1.0 \text{ kg hay as fed provides } - 0.85 \times 8.4 = 7.1 \text{ ME}$$
$$1.0 \text{ kg silage as fed provides } - 0.25 \times 9.0 = 2.25 \text{ ME}$$

Obviously it would take over 3 kg of this silage to provide the same ME as in 1 kg hay.

In practice rations for stock are calculated in terms of DM and it is only necessary to convert the DM quantities back to feeds-as-fed in the final stage of the calculation.

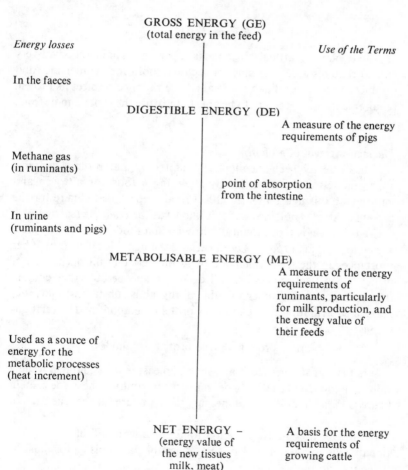

GROSS ENERGY (GE)
(total energy in the feed)

Energy losses

Use of the Terms

In the faeces

DIGESTIBLE ENERGY (DE)

A measure of the energy
requirements of pigs

Methane gas
(in ruminants)

point of absorption
from the intestine

In urine
(ruminants and pigs)

METABOLISABLE ENERGY (ME)

A measure of the energy
requirements of
ruminants, particularly
for milk production, and
the energy value of
their feeds

Used as a source of
energy for the
metabolic processes
(heat increment)

NET ENERGY –
(energy value of
the new tissues
milk, meat)

A basis for the energy
requirements of
growing cattle

"M over D" (M/D): When a ration is made up of a number of feeds the ME value of the ration as a whole is an average of the ME contributed by each of the individual feeds. The ME value of the ration as a whole is known as the "M over D" (M/D) of the ration. It is simply the total ME of the ration (M) divided by the total DM (D), and is in effect a measure of the concentration of the ration. This is of particular significance in the rationing of growing cattle and sheep.

PROTEIN

Protein plays a particularly important part in the animal body. Every animal requires a definite amount each day both for maintenance and production. In order that an animal can be rationed to meet its protein requirements it is necessary to be able to measure the protein in its food.

The measurement of protein

It has already been mentioned (Chapter 5) that protein contains approximately 16% nitrogen. By analysing a food for nitrogen and multiplying this figure by 6.25 (100/16), an approximate figure for the protein content is obtained and it is known as the *crude protein* figure.

Crude protein is the amount of nitrogen in a food multiplied by 6.25.

The term crude protein is used because the figure is affected by other substances in the feed which, although they contain nitrogen, are not proteins. Such substances are called **amides**, and because they contain free amino acids and other valuable nitrogenous substances, they, too, make a contribution to the animal's protein metabolism. The relationship is:

$$\text{crude protein} = \text{true protein} + \text{amides}.$$

What is actually needed is an accurate measure of the **protein feeding value** of a food, and as the amides make a contribution the crude protein figure is in fact a more useful measure than a figure for the true protein alone.

No matter how much protein a food contains it will have no nutritional value to an animal if it is not digestible, and this factor must be taken into account.

Digestibility is the amount of a food or nutrient that undergoes digestion and is utilised by the animal during its passage through the intestine.

Having analysed the food for protein it must then be submitted to a digestibility trial. In essence this entails feeding a known quantity of the food to an animal and then analysing the faeces and urine to determine what proportion has been wasted. By deduction the amount digested can be calculated.

It would be impossible for an agricultural chemist to submit each food he has to analyse to a digestibility trial, but from a knowledge of the results of previous trials on similar feeds he can quickly calculate the **digestible crude protein** (DCP) from the crude protein figure.

If a sample of hay or silage is sent for analysis it is the DCP figure that is contained in the report and it is a measure of the protein feeding value of the feed. The average DCP values for different feeds can be obtained from feed analysis tables (see page 68) and is expressed as so many grams (g) per kilogram (kg) of the dry matter (DM) of the feed.

Barley, for example, has a DCP of 82 g/kg DM, or, simply,

a DCP of 82
or 82 DCP.

Care must be taken when reading the protein analysis on a bag of proprietary concentrate feed as it is the crude protein figure which is normally used, e.g.:

16% protein—on a bag of dairy nuts.

For concentrates of this type the DCP is about 80% of the crude protein figure and the 16% becomes 12.8% DCP:

$$16 \times 0.8 = 12.8.$$

A further complication is that the 16% refers to 16% of the feed **as fed** and not to 16% of the DM. Likewise the 12.8% DCP is 128 g/kg of the food as fed. Such a compound has a DM of 860/kg, i.e. 86%, and so the DCP in the DM of this feed is 149 g, i.e.

$$\frac{128 \text{ g}}{0.86} = 149 \text{ g}.$$

Table 3 gives the four ways of expressing the protein content (protein feeding value) and is useful in transposing the analysis given for compound feeds into that given in the feed analysis tables.

TABLE 3

% crude protein (CP) in feed as fed	% DCP in feed as fed (CP × 0·8)	DCP g/kg in feed as fed (% DCP × 10)	DCP g/kg DM (DCP g/kg as fed × 0.86)
10	8	80	93
12	9.6	96	112
14	11.2	112	130
16	12.8	128	149
18	14.4	144	167
20	16.0	160	186

The important point is that when one feed is being compared with another the same figure is applied to each. It is particularly important when costly high protein feeds are being bought on a protein content basis.

An animal's daily protein requirements are also expressed in terms of grams of DCP. An average size Friesian cow, for example, needs 330 g DCP for maintenance and a further 51 g for every kg of milk it produces (see Chapter 11).

FEEDING STANDARDS

Standard figures have been calculated for the protein and energy requirements of all classes of stock for maintenance and for each corresponding unit of production. Appetite capacity in terms of dry matter intake are also known. These figures are known as the **feeding standards.**

All animals are individuals and some utilise their feed efficiently, and the standards laid down are adequate. Others, probably because of different temperament, are not as efficient converters of food into milk or meat and require more feed to achieve the same results. Although there are **feeding standards** there are few **standard animals.**

The feeding standards must be regarded as a **guide** to any individual animal's needs. They are much more applicable when a group of animals is being considered, and being **average** figures they must be applied to sufficient numbers of stock to give an expected **average** result.

TO SUMMARISE

The major nutritional requirements of an animal are a supply of energy and a supply of protein. To meet these needs it is necessary to have an accurate measure of how much of each any feed can supply. The figure widely used for the measurement of the protein feeding value of a feed is the digestible crude protein figure (g DCP/kg DM). A measure of the energy value is known as the metabolisable energy (ME), and all the major constituents of the food, including the protein, contribute to this. The metabolisable energy is measured in megajoules (MJ) and expressed as MJ/kg DM. The term M/D is usually applied to a ration as a whole and is a measure of the overall energy concentration.

QUESTIONS

1. Give a definition of the following terms:
   (a) crude protein;
   (b) amides;
   (c) digestibility;
   (d) DCP;
   (e) metabolisable energy;
   (f) DM and DMI;
   (g) feed as fed.
2. What do you understand by the terms:
   (a) maintenance allowance;
   (b) production allowance.
3. What is the significance of DMI?
4. What do you understand by the term "feeding standards"?
5. What are the limitations of "feeding standards" when planning a ration?

PRACTICAL

1. From feed analysis tables list the DM, ME, and DCP for the following foods:
   (a) grass silage—made in a clamp and in a tower;
   (b) hay from grass with high digestibility;

  (c) hay from grass with low digestibility;
  (d) barley and oats;
  (e) beans and peas;
  (f) decorticated groundnut cake;
  (g) linseed cake;
  (h) palm-kernel cake.
2. Compare the price per tonne of the items in 1(d) and (f) above (see the current prices in the market section of the weekly farming papers).
3. Find a bag in which pig or poultry meal has been delivered and see what analysis is given on it. Which protein figure has been used?
4. Turn to the market section of one of the weekly farming papers. In the feedingstuffs section under "oil cakes and meals", note the percentages quoted alongside each item. To what do these refer? How do these figures compare with the standard figures in the feed analysis tables?
5. Calculate the daily DMI of one of your cows or beef animals working from a typical daily ration. How does this compare with the theoretical standards? (See Chapter 11, Table 9, and Chapter 12, Table 17.)

Feeding dairy cattle for milk production

WINTER FEEDING

During the winter dairy cattle are normally housed, and it is possible to have a large measure of control over their feeding. Farm-grown fodder, including such feeds as hay, kale, roots, silage, barley, and purchased concentrates, provides the cows with the nutrients they need to keep healthy and to produce milk and calves. It is important that, from the feeds he has available, the farmer draws up a ration that meets the cow's requirements as accurately as possible, particularly in respect of energy and protein. The prime consideration is not only to supply sufficient of each, but to make sure that in meeting the need for one the ration does not have a wasteful surplus of the other. Minerals and vitamins are equally important but here the problem is not so great. Any shortage can easily be rectified by adding small amounts of a relatively cheap mineral/vitamin supplement.

THE METABOLISABLE ENERGY (ME) SYSTEM

There is more than one way of arriving at the correct quantity and correct balance of energy and protein in a ration, but the present system, which is being advocated by the agricultural advisory services, is known as the **metabolisable energy system**. The basis of this system is to calculate the total energy requirements for the cows from their requirements for:

(a) maintenance;
(b) milk production;
(c) an adjustment for liveweight gain or loss.

Requirements for protein, minerals, and vitamins are calculated separately.

Maintenance

The maintenance needs of a dairy cow vary with the size of the animal —the larger the animal the greater the requirements—and this is shown in the recommended feeding standards (for both energy and protein).

TABLE 4.

Daily maintenance requirements for dairy cows according to liveweight and breed

Breed	Liveweight (kg)	Maintenance allowance	
		Energy (MJ of ME)	Protein (g/DCP)
Jersey	360	41	231
Ayrshire	500	54	295
Friesian	600	63	331

From Table 4 it can be seen that the daily maintenance allowance for an average size Friesian cow should contain 63 ME, 331 DCP.

These nutrients are contained in about 8 kg of good quality hay. On the other hand, the maintenance of a Jersey cow would be met by only 6 kg of the same hay.

Production

The cow's feeding requirements for each kilogram of milk produced vary with the quality of the milk. The allowances for energy and protein for the milk from the three breeds, based on the average butterfat (BF) and solids-not-fat (SNF), are given in Table 5.

Liveweight Change Allowance

The third consideration in rationing dairy cows is to make an allowance for the changes in bodyweight during the production cycle. It is well

TABLE 5.

Allowances of energy and protein for milk production

Breed	Milk composition		Allowance 1 kg milk	
	BF (g/kg)	SNF (g/kg)	MJ of ME	g DCP
Jersey	48	91	5.9	63
Ayrshire	37	88	5.1	53
Friesian	35	86	4.9	51
Average milk	36	86	5.0	51

known that the milk yield of a cow builds up to a peak some 6–8 weeks after calving then slowly diminishes as the lactation progresses. During these early weeks, although the appetite also increases, the cow is not able to take in sufficient feed to meet her total requirements. In early lactation, therefore, she tends to "milk off her back" and loses weight.

During mid-lactation she usually holds her own without further loss, and in late lactation and through into the dry period she is able to regain the lost weight.

Each kilogram of liveweight loss is equivalent to 28 MJ of dietary energy. A high yielding cow may lose 0.5 kg/day and thereby contribute 14 ME to her total energy requirement.

When the cow is replacing her lost liveweight an extra 34 ME must be provided in the diet for every kilogram of tissue replaced. From 20 weeks after calving to the end of lactation a gain of 0.5 kg/day would be satisfactory and in the dry period 0.75 kg/day.

The desirable liveweight changes and adjustments to the total ration throughout the production cycle are summarised in Table 6.

TABLE 6.

Desirable liveweight changes and dietary adjustments

Weeks after calving	Liveweight change (+ or − kg/day)	Adjustment to total maintenance and production requirements (+ or − MJ)
0–10	−0.5	−14
10–20	—	—
20–40	+0.5	+17
40–52	+0.75	+25

Calculation of Total Energy and Protein Requirements

From the data given in Tables 4–6 it is now possible to calculate the energy (ME) and the protein (DCP) required by a cow in any of the three periods of her lactation.

Example A

A Friesian cow (600 kg liveweight) in early lactation giving 25 kg milk (35 g/kg BF, 86 g/kg SNF) will require:

	ME	DCP
For maintenance (Table 4)	62	331
For production (Table 5)		
(4.9 ME and 51 DCP per kg milk) × 25	122	1275
	184	1606
Adjustment for bodyweight change (Table 6)		
0–10 weeks after calving		
0.5 loss of weight which contributes	−14	
Total requirement from the food	170 ME	1606 DCP

Table 7 combines all three factors—size of cow, stage of lactation, and quantity/quality of milk into one table for the daily ME allowances. Likewise, Table 8 does the same for the DCP allowances.

Degradable and undegradable protein

The protein taken in by a cow (a ruminant) becomes available for milk production and tissue growth in two ways. In the first place it may be broken down (**degraded**) by the bacteria in the rumen and used in their own protein synthesis. Later in the digestion process the bacteria are themselves digested and the protein they contain becomes available to the animal for its own protein synthesis. On the other hand, some protein is resistant to bacterial breakdown (**undegradable**) and passes unchanged into the small intestine where it is digested and absorbed in the usual way.

At a low level of productivity a cow can meet her protein needs from

TABLE 7.

Daily ME allowances for three breeds of dairy cattle (MJ/day)

Breed	Liveweight change	Main-tenance	5	10	15	20	25	30	35
						Milk yield (kg/day)			
FRIESIAN									
(600 kg)	Losing 0.5 kg/day	—	73	97	112	147	172	196	221
36 g/kg BF	No weight change	63	87	111	136	161	186	210	235
86 g/kg SNF	Gaining 0.5 kg/day	—	104	128	153	178	203	227	252
AYRSHIRE									
(500 kg)	Losing 0.5 kg/day	—	66	92	118	144	169	195	
38 g/kg BF	No weight change	54	80	106	132	158	183	209	
89 g/kg SNF	Gaining 0.5 kg/day	—	97	123	149	175	200	226	
JERSEY									
(360 kg)	Losing 0.5 kg/day	—	58	88	118	149	180		
49 g/kg BF	No weight change	41	72	102	132	163	194		
95 g/kg SNF	Gaining 0.5 kg/day	—	89	119	149	180	211		

Including safety margin.

TABLE 8.

Allowances for digestible crude protein (DCP) for dairy cows (g/day)

Breed			Milk yield (kg/day)				
	0	10	15	20	25	30	35
Friesian	331	846	1103	1361	1618	1875	2132
Ayrshire	295	829	1096	1363	1631	1898	—
Jersey	231	863	1178	1494	1810	—	

From these tables the requirements for the cow in Example A are rounded at 172 ME and 1618 DCP.

the microbial protein, i.e. that provided by digesting the bacteria. In these circumstances, therefore, the diet only needs to contain a supply of degradable protein. High-yielding cows, however, cannot meet all their protein needs from that supplied by the bacteria and need a proportion of undegradable protein in the diet.

The protein in silage and barley is mainly degradable (85%), and, therefore, supplies only a small amount of undegradable protein. Some high protein concentrates such as soyabean meal, and especially fish meal, contain a higher proportion of undegradable protein, and these feeds should be used as protein supplements in the concentrate mixes of high-yielding cows. Some manufacturers use this as a sales feature for their compound feeds and high protein supplements.

The feeding standards for dairy cows and the protein content of feeds may, in the future, be given in these terms. For the purpose of this text no attempt is made to apply this advanced system of protein rationing apart from the practical suggestions mentioned above.

APPETITE

Of far more importance is the feed intake of the cow which is basic to milk production. The greater the intake the greater the milk yield potential. Feed intake or appetite is expressed in terms of dry matter intake (DMI), and there are two main factors which affect it:

(a) the size of the cow—from 2 to 3% of bodyweight;
(b) the M/D of the ration—the higher the M/D the greater the intake.

TABLE 9A

M/D of the ration	DMI as a percentage of bodyweight	For a 600-kg cow (kg DMI)
10.0	2.0	12.0
10.5	2.2	13.5
11.0	2.4	15.0
11.5	2.7	16.5
12.0	3.0	18.0
12.5	3.3	19.5

Table 9B expands these factors giving the total ME intake in each case. For example, a 600-kg cow on a ration with an M/D of 11.0 has a DMI of 15.0 giving an ME intake of 165 (11 × 15.0).

More importantly, the table can be used in reverse when formulating a ration to meet a calculated ME requirement. The ME allowance for a cow weighing 500 kg, for example, may have been calculated as 155 ME. The table shows that the minimum M/D would need to be 11.5 and the cow would have a DMI of 13.5 kg. A suitable combination of bulk and concentrate feeds could then be devised to fall within these limits.

TABLE 9B.

DMI and ME intake related to M/D of the ration and size of cow

M/D of ration	650-kg cow		600-kg cow		550-kg cow		500-kg cow		400-kg cow	
	DMI	ME	DMI	ME	DMI	ME	DMI	ME	DMI	ME
10.0	13.0	130	12.0	120	11.0	110	10.0	100	8.0	80
10.1	13.3	134	12.3	124	11.1	112	10.1	102	8.2	83
10.2	13.7	140	12.6	128	11.2	114	10.3	105	8.4	86
10.3	14.0	144	12.9	133	11.5	118	10.5	108	8.5	88
10.4	14.3	149	13.2	137	11.8	122	10.8	112	8.6	89
10.5	14.6	153	13.5	142	12.1	127	11.0	116	8.7	91
10.6	14.9	157	13.8	146	12.4	131	11.3	119	8.8	93
10.7	15.0	160	14.1	151	12.7	135	11.5	123	9.0	96
10.8	15.3	165	14.4	156	13.0	140	11.8	126	9.2	99
10.9	15.6	170	14.7	160	13.3	145	12.0	130	9.4	102
11.0	15.9	175	15.0	165	13.6	150	12.2	134	9.6	106
11.1	16.2	180	15.3	170	13.9	154	12.5	138	9.8	109
11.2	16.5	185	15.6	175	14.2	159	12.8	143	10.0	112
11.3	16.8	190	15.9	180	14.5	163	13.0	147	10.3	116
11.4	17.1	195	16.2	185	14.8	168	13.3	151	10.5	120
11.5	17.5	201	16.5	190	15.1	174	13.5	155	10.8	124
11.6	17.9	207	16.8	195	15.4	179	13.8	159	11.0	128
11.7	18.3	214	17.1	200	15.7	184	14.1	164	11.2	131
11.8	18.7	221	17.4	205	16.0	189	14.5	170	11.4	134
11.9	19.1	228	17.7	211	16.3	194	14.8	175	11.6	138
12.0	19.5	234	18.0	216	16.5	198	15.0	180	11.8	142
12.1	19.9	241	18.3	221	16.8	204	15.3	185	12.0	146
12.2	20.3	248	18.6	227	17.1	209	15.5	190	12.3	150
12.3	20.7	255	18.9	232	17.4	214	15.8	195	12.6	155
12.4	21.1	262	19.2	238	17.7	219	16.0	200	12.9	160
12.5	22.7	284	19.5	244	18.0	225	16.5	206	13.2	165

COMPILING A RATION FOR A LACTATING COW

A ration is required for a cow weighing 600 kg (Friesian) giving 30 kg milk in the eighth week of lactation. The feeds available are good quality clamp silage (250 DM, 10.5 ME, 105 DCP), barley, and a grain balancer cake (880 DM, 12.5 ME, 280 DCP).

Step 1. Determine the energy and protein needs of the cow.

     ME (from Table 7)   196
     DCP (from Table 8)   1875

Step 2. Determine the analysis of the feeds.

	ME	DCP	DM
Silage (actual)	10.5	105	250
Barley (tables)	13.0	82	850
Grain balancer (actual)	12.5	280	880

Step 3. Determine the quantities of silage and barley mix* that will supply 196 ME within the appetite limits of the cow.

> From Table 9: 196 ME for a 600-kg cow requires an M/D of 11.6 giving a DMI of 16.8 kg.

The proportion of DM silage 10.5 ME and DM barley mix 13.0 ME to give an overall M/D of 11.6 can be calculated by using the **energy balancing chart** (Fig. 2). The proportions are as follows:

55% silage	9.2 kg DM silage	36.8 kg silage as fed
OR		OR
45% barley mix	7.6 kg DM barley mix	8.8 kg barley mix as fed

$$9.2 \text{ kg DM silage} \times 10.5 = 97$$
$$7.6 \text{ kg DM barley mix} \times 13.0 = 99$$

$$\overline{196 \text{ ME}}$$

Step 4. Determine the quantity of protein supplied by the silage and the balance to be made up form the barley mix.

Total DCP required	1875
DCP in 9.2 kg silage ($\times$ 105)	966
DCP to be supplied by barley mix	909 in 7.6 kg of the mix

---

*The exact proportions of barley and grain balancer are not yet known but at this point an ME value of 13.0 can be assumed.

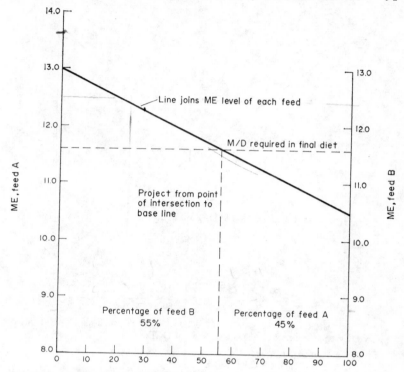

FIG. 2. ENERGY BALANCING CHART. To determine the proportion of two feeds of known ME to give a final diet of a specified ME, i.e. specified M/D.

From the example on p. 90:
Feed A: 13.0 ME,
Feed B: 10.5 ME.   M/D required in final diet, 11.6.

The protein level of the mix must be

$$\frac{909}{7.6} = 120 \text{ g DCP/kg DM of the mix}$$

Step 5. Calculate the proportions of barley and grain balancer in the mix to give this level of protein.

The quantities of barley (82 DCP) and grain balancer (280 DCP) to give a mix of 120 DCP can easily be determined by

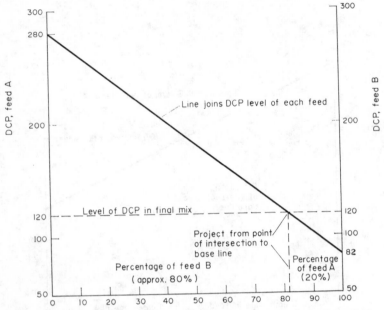

FIG. 3. PROTEIN BALANCING CHART. To determine the proportions of two feeds of know DCP to give a mix of a specified level of DCP.

From the example of p. 91:
Feed A: grain balancer cake, 280 DCP,
Feed B: barley, 82 DCP.   Level required in final mix, 120 DCP.

using the **protein balancing chart** (Fig. 3). The proportions are as follows:

80% barley and 20% grain balancer, which can be checked—giving an analysis for the mix of **12.9 ME   121 DCP**. The complete ration then becomes as shown in Table 11.

TABLE 10

	Analysis of DM		ME	DCP
8.0 kg DM barley	× 13.0	=	104	
	× 82	=		656
2.0 kg DM grain balancer	× 12.5	=	25	
	× 280	=		560
			129	1216

TABLE 11.

Feed as fed	kg DM		Analysis of DM		ME	DCP
36.8 kg silage	9.2	×	10.5	=	97	
		×	105	=		966
8.8 kg barley mix	7.6	×	12.9	=	98	
		×	121	=		920
	16.8 kg DM		providing		**195 ME**	**1886 DCP**

Minimum fibre content

There is a minimum fibre content below which the ration should not fall. If it does, digestive upsets are likely to develop, and the quality of the milk may also be impaired. The minimum fibre content appears to be about 15% or 150 g/kg DM for the ration overall.

Silage contains around 30% crude fibre but barley has only 5.3%. As a general rule the concentrate element should not be more than 60% of the total DM. If this level has to be exceeded, then it becomes necessary to raise the percentage of fibre in the concentrate mix. Molassed sugar-beet pulp has a fibre content of 12% and can be used to replace some of the barley.

The problem underlines the importance of high-quality silage for high-yielding cows, thereby reducing the need for such a high proportion of concentrates in the ration. In the example above, had the ME of the silage been 10.0 instead of 10.5, the proportion of concentrates would have to have been 55% instead of 45%—nearing the 60% limit.

RATIONING FOR THE COMPLETE PRODUCTION CYCLE

If the ME system was to be taken literally, for any one cow the ration would be changing daily in terms of bulk feed to concentrate and the balance of protein within the concentrate. This is a non-starter even for a single cow, let alone a herd. It is a fallacy to believe that feeding dairy cows is an exact science.

Cows within a herd should be grouped for feeding and the best basis for grouping is on calving date. At least all the members of the group will be at the same point in the lactation. The average yield of the group will be known from milk recording and previous lactation records, and from these it is possible to draw up an average lactation curve. This has been done in Table 12 for an imaginary herd of: **600-kg cows—calving around 1 October—with a 6000-kg lactation yield—peaking at 30 kg.**

TABLE 12

Month	Approximate daily yield (kg)	Nutritional needs	
		ME	DCP
October	25	172	1618
November	30 (peak)	196	1875
December	27	182	1720
January	25	186	1618
February	22	171	1463
March	20	178	1361
April	17	163	1205
May	14	148	1052
June	12	138	948
July	10	136	846
August	—	87	455
September	—	87	595

Table 12 is straightforward and is equivalent to **Step 1** in the previous example. It assumes a loss of weight of 0.5 kg/day from October to December, a static situation in January and February, an increase of 0.5 kg from March to June, and 0.75 kg from July to September.

The feeds available are given in Table 13. The figures for silage barley and grain balancer are as before. Two values for grass have been taken to represent the fall in value during the summer months.

TABLE 13

Period	Feed	ME	DCP	DM
October to April	Silage	10.5	105	250
	Barley	13.0	82	850
	Grain balancer	12.5	280	880
May and June	Grass (close grazing)	12.0	185	200
July to September	Grass (extensive grazing)	10.5	130	200

Table 14 uses Table 9 to determine the minimum M/D and the corresponding DMI to meet the ME of the ration in each month. The energy balancing chart (Fig. 2) has then been used to calculate the quantities of silage and barley mix to provide the ME within the DMI. From May to September the cows are at grass, and amounts of DM grass are given that will provide the ME required. Assuming the grass is readily available there should be no problem of the cows consuming these amounts of dry matter. The 11.8 kg in May is below 2% of the cow's bodyweight.

TABLE 14

Month	ME	M/D	DMI	Silage kg DM (ME)		Barley mix*(a) kg DM (ME)	
October	172	11.1	15.3	11.4	(120)	3.9	(51)
November	196	11.6	16.8	9.2	(97)	7.6	(99)
December	182	11.3	15.9	10.3	(108)	5.6	(73)
January	186	11.4	16.2	9.7	(102)	6.5	(84)
February	171	11.1	15.3	11.4	(120)	3.9	(51)
March	178	11.3	15.9	10.3	(108)	5.6	(73)
April	163	11.0	15.0	11.75	(123)	3.25	(42)
				grass			
May	148	10.6	13.8	11.8	(148)	—	
June	138	10.4	13.2	11.0	(138)	—	
July	136	10.4	13.2	10.9	(136)	—	
August	87	—	12.0	8.3	(87)	?	
September	87	—	12.0	8.3	(87)	?	

*(a)ME of 13.0 assumed for the mix.

If all has gone to plan, the cows should be dry in August and September, and, providing there is sufficient grass available for them to take in 8–9 kg DM per day, this will meet their needs. This is, nevertheless, a considerable amount of herbage—about 70% of what might be consumed at the height of the grazing season, and in all probability consumption may not be high enough. Normal practice is to commence feeding concentrates in the last 6 or 8 weeks before calving, say from 1·0 kg rising to 4·0 kg/day for cows of this level of production. Such feeding also gives the opportunity to build up the mineral reserves and reach a level of concentrate feeding about 75% of the level which will be fed after calving.

The final step is to determine the level of protein in the barley mix. This is given in Table 15 and utilises the protein balancing chart.

TABLE 15

Month	DCP in the silage	DCP to be supplied by the mix	DCP level in the mix (g/kg DM)	% grain balancer
October	1198	420 in 3.9 kg	108 ⎫	
November	966	909 in 7.6 kg	120 ⎬	20 (123DCP)
December	1081	639 in 5.6 kg	114 ⎭	
January	1018	600 in 6.5 kg	92	10 (100DCP)
February	1198	265 in 3.9 kg	70 ⎫	nil (82DCP)
March	1081	280 in 5.6 kg	50 ⎭	
April	1233 in the grass	Nil	—	—
May	2183			
June	2035			
July	1417			
August	1079			5[a] (90DCP)
September	1079			10[a] (100DCP)

[a]Assuming concentrates are fed.

Over the first 3 months of the lactation the level of protein in the mix is fairly constant. A 20% inclusion rate for the grain balancer will easily cover the protein requirements for this period. Likewise in January the 10% inclusion more than meets the requirement for a level of 92. From February onwards no protein supplementation is required at all as the required level of 70 is below the level in barley (82).

When concentrates are fed in the dry period (for steaming up) a protein supplement should be added in order to ensure the protein needs are met and to reintroduce the cows to the diet they will receive after calving.

It must be remembered that the above figures only apply to this example where the bulk feed is silage with a fairly high protein content.

Minerals

The mineral requirements of high-yielding cows are substantial, and an inadequate supply can be a limiting factor on yield. For mainten-ance an average cow needs:

20 g calcium (Ca), 30 g phosphorus (P), 12 g magnesium (Mg), and 10 g sodium (Na);

and for each kilogram of milk produced:

2.8g Ca; 1.7g P; 0.6g Mg; 0.6g Na.

Table 16 shows the need of the high-yielding cow (30 kg peak yield) in the example.

TABLE 16

	Ca	P	Mg	Na
Maintenance	20	30	12	10
Production	84	51	18	18
	104	81	30	28
The ration quoted provides:				
9.2 kg silage	51	26	12	34
7.6 kg barley mix	4	30	10	1
	55	56	22	35

As it stands the ration is deficient of 49 g Ca, 25 g P, and 10 g Mg, although no allowance has been made for the mineral content of the grain-balancer supplement. 0.18 kg of steamed bone flour would rectify the deficiency of calcium and phosphorus and 0.02 kg of calcined magnesite the deficiency in magnesium. This is a little over a 2% inclusion rate and it is usual to **add up to 2% minerals to a concentrate mix.**

It should be noted how the silage and barley mix complement each other for calcium and phosphorus. If the ration only consisted of silage, phosphorus could be in short supply. The magnesium content could also be dangerously low. Such a possibility could occur in the latter part of the lactation when a greater proportion of the ration is silage; and, of course, when the cows go out to grass.

Young grass is high in calcium, and particular care has to be taken when the cows are turned out in the Spring. The sudden increase in dietary calcium can upset the calcium/magnesium balance—particularly if turnout is associated with a lower level of concentrate feeding. Calcined magnesite can be fed (50 g/day in 1 kg barley) to offset this problem. It should be fed from before turnout to at least 2 weeks after.

Vitamins

Under normal conditions when a variety of reasonable quality feeds are included in the ration it is not considered necessary to include a vitamin supplement. Furthermore, ruminants are able to synthesise vitamins of the B complex and do not have to rely on an external supply. When feed supplies are of poor quality, vitamins A and D_3 are sometimes included in winter rations mainly as a general insurance.

Summer feeding

There are several factors which make the summer rationing of dairy cows more difficult than winter feeding. Most of the difficulties arise because the cattle are out at grass. In the first place grass is a very variable product yielding different proportions of metabolisable energy and protein at different stages of maturity. Secondly, it is extremely difficult to make an accurate estimate of the animals' daily consumption, which varies with the height of the grass, the density of the sward, and the overall palatability. The same factors also effect the amount of energy an animal expends on walking in search of its food.

One of the aims of good grassland management is to utilise the crop at its most nutritious stage of growth and to endeavour to maintain an ample amount of highly digestible herbage always before the cows. To achieve this, some form of controlled grazing is usually practised either by an electric fence or paddock system. By such methods the sward can be fully grazed and rested for 3 or 4 weeks before a second crop of grass some 15–30 cm high is available for the next grazing.

Many of the variants are then eliminated as the cows have a good quantity of high-quality feed always before them. Grass in this early stage of growth is highly digestible (has a high D-value), cows expend little energy on consuming it and the daily consumption is high. The only variant that cannot be controlled is the weather. Under wild wet conditions consumption is likely to be reduced considerably and the ME system of rationing relies on the animals feeding to the capacity of their appetite. An extra allowance of concentrate or hay may be needed under such conditions. No doubt the only fully controlled system of summer rationing is zero grazing, when the cows are yarded and grass carted to them.

An average-size cow under good grazing conditions may consume up to 70 kg of fresh grass per day containing some 14 kg DM. The analysis of grass grazed rotationally at 3-week intervals is:

$$12.1 \text{ ME}, \quad 183 \text{ DCP}.$$

The daily intake would therefore be:

$$169.4 \text{ ME}, \quad 1562 \text{ DCP},$$

—which would cater for the total needs of most cows giving up to 20 kg milk per day whatever point they were in the lactation. There would in fact be a surplus of protein and, therefore, cows yielding more than this would need to have a supplement of a high energy concentrate such as barley at, say, 0.4 kg as fed/kg milk over 20 kg.

A lack of fibre may also be a problem during the period of lush growth. It has been shown in practice that even high-yielding cows benefit from a feed of 1–2 kg of hay or straw before turning out on to young, rapidly growing grass. The fibrous feed is eaten readily, and both composition and yield of milk are improved.

As the season progresses the D-value of the grass tends to deteriorate. In late May and early June the grasses go to seed even when closely grazed.

The weather is also drier and the combination of these factors reduces the quantity and quality of the herbage. The effects are often more marked in the drier parts of the country than in the regions where the summer rainfall is higher. Some swards become unpalatable by continual soiling with dung and urine, which further reduces consumption.

With the autumn rains there is a second flush of grass in late August and September. At this time of year the quality of the grass never seems to be as high as in the spring. The analysis, however, does not reflect this apparent difference, and it is probably the physical condition of the herbage which reduces the intake. Heavy dews and wet weather in the autumn may be the cause of reduced consumption.

Throughout the late summer and early autumn there is a problem, therefore, of not knowing how much the bulk feed, the grazing, is contributing to the nutrition of the cow. For an autumn-calving herd the cows will either be in late lactation or in the dry period, and the important point is to see that they are improving in condition to regain the weight they lost earlier in the lactation. This they may well do without any additional feed, but if the grass is getting scarce a decision has to be made as to how much reliance can be placed upon it. Supplementation with a concentrate as calving approaches may be all that is necessary, and, indeed, this is good practice if only to accustom the cow to the diet it will be having after calving. On the other hand, there may also be a period when additional bulk feed, such as hay, should be made available. Eventually it becomes necessary to move on to full winter rationing.

Spring calvers could still be yielding well as autumn approaches, and high yielders may have been receiving some supplementary feed throughout. To avoid any loss in yield it may be necessary to bring them on to full winter rations sooner than the dry cows, or at least supplement the grazing with hay or other bulk feed. As a general rule it should not be assumed that autumn grass provides much more nutrition than that needed for maintenance.

There can be no hard and fast rules, however, for this difficult period, when the aim is to make the most of the grazing available but without prejudicing the yield or condition of the cows. A sudden change in the weather can radically alter the situation overnight.

QUESTIONS

1. What three factors are taken into consideration when calculating a cow's energy needs?
2. In what way does the size of the cow and the quality of the milk affect the calculation?
3. What is the daily ME allowance for a Friesian?
4. What quantity of good hay would supply these nutrients?
5. Over what period of the lactation does a cow (a) lose weight; (b) hold her own; (c) gain weight?
6. What do you understand by the terms **degradable** and **undegradable** protein? What is the significance for high-yielding cows?
7. Determine the ME and DCP allowances for the following cows:
    (a) a Friesian in mid lactation giving 20 kg milk per day;
    (b) a Jersey 5 weeks after calving giving 15 kg per day;
    (c) an Ayrshire in late lactation giving 10 kg milk per day.

8. From Table 9 determine the M/D of the ration for the above cows and the DMI.

9. Using the energy balancing chart (Fig. 2), determine the percentages of the following pairs of feeds to give the required M/D:
   - (a) hay 8.2 ME and barley mix 13.0 ME to give an M/D of 11.0;
   - (b) clamp silage 10.1 ME and a compound cake 12.0 ME to give an M/D of 10.5;
   - (c) tower silage 10.5 ME and a 50/50 mixture of oats 11.5 ME and barley 13.0 ME to give an M/D of 11.1.

10. Assuming the cows receiving the rations in question 9 had a DMI of 16.0, calculate quantities of the feeds as fed.

11. Using the protein balancing chart (Fig. 3), determine the percentages of the following pairs of feeds to give the required level of protein:
    - (a) barley 82 DCP and groundnut cake 449 DCP to give 150 DCP;
    - (b) a mix of 1 part oats and 1 part barley (84 and 82 DCP) with soyabean meal 453 DCP to give 145 DCP:
    - (c) a mix of 1 part flaked maize 106 DCP and 3 parts barley with a high protein compound 350 DCP to give a final mix of 135 DCP.

12. If a 600-kg cow consumed 60 kg of fresh rotationally grazed grass, what level of production might be expected assuming she was in mid-lactation?

CHAPTER 12

Rations for beef cattle

ALTHOUGH many of the principles that are applied to the rationing of dairy cows also apply to beef cattle, there are at least two major differences. The first of these is the wide range of diets on which beef cattle are fed which stem from the equally wide range of systems under which they are managed. Some are managed to achieve maximum growth rate and fed almost entirely on concentrates. At the other extreme the main object may be to make the most effective use of arable byproducts or an area of rough grazing with growth rate being of secondary importance. Between these extremes there is a range of systems and diets which would not normally apply to dairy cows.

A second difference is more fundamental and relates to the way in which animals use the energy in their diet for **growth** compared with the way in which it is used for milk production. The difference is partly, physiological and partly due to the variable composition of the new tissue which makes up the liveweight gain. How these factors affect the rationing will be seen later in the chapter.

APPETITE—DMI

The appetite of a beef animal in relation to its size is somewhat less than that of a dairy cow. Lactating animals tend to be hungry by nature and increase their DMI after parturition. Like all young creatures, the young beef animal has the greatest DMI in relation to its bodyweight at about

3 kg DMI per 100 kg liveweight (3% of liveweight).

As it gets older and heavier its DMI decreases relatively to under

2 kg DMI per 100 kg liveweight (2% of liveweight).

As with the dairy cow, the DMI will also be affected by the M/D of the ration, and the figures above assume a ration with an M/D of about 11.0. At an M/D of 10.0 the feed intake will be lower and higher at an M/D of 12.0 or more.

ENERGY FOR MAINTENANCE

As might be expected, the energy allowances for maintenance are the same as for dairy cattle. Table 17 gives the DMI and the ME for maintenance according to liveweight—again an M/D of 11.0 is assumed for the DMI.

TABLE 17

*Appetite and energy allowances for mainten-
ance related to liveweight*

Liveweight (kg)	Appetite (kg DMI)[a]	ME for maintenance (MJ)
100	3.0	17
150	4.0	22
200	5.0	27
250	6.0	31
300	7.0	36
350	8.0	40
400	8.5	45
450	9.0	49
500	10.0	54
550	11.0	59
600	11.5	63

[a]Deduct 1.0 kg for a ration with an M/D
of 10.0. Add 1.0 to 2.0 kg for rations with
an M/D of 12.0 or more.

ENERGY FOR PRODUCTION

The efficiency with which an animal utilises the energy in its diet to increase its liveweight depends on two factors:

(a) the concentration of energy in the diet (M/D);

(b) the rate at which it is growing in relation to its size,
   i.e. the **animal production level (APL).**

As these two factors change so does the amount of ME that has to be fed for a particular unit of growth.

CONCENTRATION OF ENERGY IN THE DIET (M/D)

The M/D has been mentioned previously (Chapter 10). The principle is simply the higher the M/D the more efficiently the animal utilises the ME in the diet. For example, the energy in barley, 13.0 ME, is utilised more efficiently than the energy in hay, which may be as low as 8.0 ME. If the same amount of energy was fed to two animals, one in the form of barley and the other as hay, the barley-fed animal would grow faster. One reason for this is that the animal uses more energy processing the bulkier material and in extracting energy from it—with a poorer net result.

The M/D of a typical ration can be calculated as Table 18 shows.

TABLE 18 *Ration 1*

Fed as fed	kg DM		ME/kg DM		ME in ration
3.0 kg hay	2.6	×	8.8	=	22.9
10.0 kg silage	2.0	×	9.2	=	18.4
2.0 kg barley	1.7	×	13.0	=	22.1
	6.3				63.4

$$M/D = \frac{63.4}{6.3} = 10.0$$

In practice this ration would be sufficient for a 300-kg animal to grow at 0.75 kg/day. The same amount of energy could be provided in a ration consisting mainly of cereal, for example (Table 19):

TABLE 19 *Ration 2*

Feed as fed	kg DM		ME/kg DM		ME in ration
0.6 kg hay	0.51	×	8.8	=	4.5
1.0 kg flaked maize	0.9	×	15.0	=	13.5
4.0 kg barley	3.44	×	13.0	=	44.7
	4.85				62.7

$$M/D = \frac{62.7}{4.85} = 12.9.$$

The liveweight gain for the same animal on ration 2 with the higher M/D would be nearly 1·0 kg/day.

In general, therefore, the more concentrated the diet the greater is the liveweight gain (LWG) for the same amount of energy supplied. On this basis the barley/beef system is more efficient than a cereal/silage system in terms of energy utilised. The cereal/silage system in turn is more efficient than an extensive grazing system where the M/D of the grass falls below 10.0. This does not mean, of course, that any one system is more economic than the next, and the grazing system may, in fact, be the cheapest. It depends on the **cost** of each megajoule of energy as well as how efficiently it is utilised.

ANIMAL PRODUCTION LEVEL (APL)

The second factor which determines the amount of energy required for production is the animal production level (APL). There are two aspects to APL: (a) the daily liveweight gain (LWG), and (b) the size of the animal.

The faster an animal is growing the more energy it needs for each unit of LWG. The following, for example, are the ME allowances for a 350-kg animal growing at different rates:

Liveweight gain (kg/day)	0.0	0.25	0.5	0.75	1.00	1.25
ME allowances (at M/D of 12)	40	47	55	65	77	92

It can be seen that for each additional 0.25 kg LWG the **additional** ME required increases over the range from 7 to 15. The reason for this is due in part to the energy used in creating new growth but also because the more rapidly an animal grows the greater the fat content of the new tissue.

For the same reasons the APL is also related to the size of the animal; again the LWG of the larger older animal tends to contain a greater proportion of fat, i.e. the LWG contains a greater amount of **net energy (NE)** and therefore it needs more energy in the diet to achieve the same result.

In the example above, the 350-kg animal needed an extra 12 ME to improve its LWG from 0.75 to 1.00 kg/day. A smaller younger animal of 150 kg needs 8 ME to make the same change; but an older larger animal of 550 kg needs 16 ME.

Table 20 below combines all these factors, M/D of the ration, growth rate, and size of the animal to give the energy allowances for maintenance and production for growing cattle.

TABLE 20
Metabolisable energy allowances for growing cattle

Liveweight (kg)	M/D of ration (MJ/kg DM)	Liveweight gain (kg/day)						
		0.2	0.4	0.6	0.8	1.0	1.2	1.4
	9	22	27					
	10	22	26	32				
100	11	21	26	31	36			
	12	21	25	29	35			
	13	21	24	29	33	39		
	9	27	33					
	10	27	32	38				
150	11	26	31	37	43			
	12	26	30	35	41	48		
	13	26	30	34	40	46		
	9	32	39	46				
	10	32	38	44	52			
200	11	31	37	43	50	58		
	12	31	36	41	48	55	64	
	13	31	35	40	46	53	62	72
	9	37	45	53				
	10	37	43	51	59			
250	11	36	42	49	57	66		
	12	36	41	47	54	63	73	
	13	35	40	46	53	60	69	80
	9	42	50	59	69			
	10	42	49	57	66	77		
300	11	41	48	55	63	73	85	
	12	41	47	53	61	70	81	94
	13	40	46	52	59	67	77	89
	9	48	56	65	78			
	10	47	54	62	73	85		
350	11	46	53	61	70	81	93	
	12	46	52	59	68	77	89	103
	13	45	51	58	65	74	85	98
	9	53	62	72	84			
	10	52	60	69	80	92		

Table 20—*continued*

Liveweight (kg)	M/D of ration (MJ/kg DM)	Liveweight gain (kg/day)						
		0.2	0.4	0.6	0.8	1.0	1.2	1.4
400	11	51	59	67	77	88	102	
	12	51	57	65	74	84	97	112
	13	50	56	64	72	81	93	107
	9	58	67	78	91			
	10	57	66	75	87	100		
450	11	56	64	73	83	96	110	
	12	56	63	71	81	92	105	121
	13	55	62	69	78	88	101	115
	9	63	73	85	98			
	10	62	71	82	94	108		
500	11	61	69	79	90	103	118	
	12	61	68	77	87	99	113	130
	13	60	67	75	85	95	108	124
	9	68	79	91	105			
	10	67	77	88	101	116		
550	11	66	75	85	97	111	127	
	12	66	74	83	94	106	121	139
	13	65	72	81	91	103	116	133

CALCULATING RATIONS FOR GROWING CATTLE

Rations can be calculated for growing cattle by basing them on either:

(a) the metabolisable energy (ME) requirements; or
(b) the net energy (HE) requirements.

Both methods are based on the principles of feed utilisation already explained and Table 20 is the basis of the ME system. Table 20 has its limitations, however, for although it is very useful in forecasting the likely performance from a given ration, it is virtually impossible to compile a ration from feeds of known analysis for an animal of a certain size to give a required LWG. In substituting one feed for another the M/D of the ration invariably changes, and thereby the efficiency of utilisation of the ME and the result does not turn out as expected.

The only means of compiling a ration from such a standpoint is through the **net energy** method. The system is based on the net energy (NE) requirements of animals according to their size and target LWG Table 21 is, therefore, the starting point.

TABLE 21

Net energy allowances for growing cattle (MJ/day)

Liveweight	Liveweight gain (kg/day)							
	0.2	0.4	0.6	0.8	1.0	1.2	1.4	1.6
100	14.2	16.3	18.6	21.4	24.6	28.4	33.0	38.7
150	17.6	19.9	22.6	25.6	29.2	33.5	38.6	45.0
200	21.0	23.6	26.5	29.9	33.8	38.5	44.2	51.2
250	24.4	27.2	30.4	34.1	38.4	43.6	49.8	57.4
300	27.8	30.9	34.3	38.3	43.0	48.6	55.4	63.7
350	31.2	34.5	38.3	42.6	47.7	53.7	61.0	69.9
400	34.7	38.2	42.2	46.8	52.3	58.7	66.5	76.2
450	38.1	41.8	46.1	51.1	56.9	63.8	72.1	82.4
500	41.5	45.5	50.0	55.3	61.5	68.8	77.7	88.6
550	44.9	49.1	53.9	59.6	66.1	73.9	83.3	94.9
600	48.3	52.8	57.9	63.8	70.7	79.0	88.9	101.1

From Table 21 it is a simple matter to read off the net energy allowance for an animal of a particular size required to grow at a certain rate.

The next step is to determine the amount of metabolisable energy needed to supply the net energy requirements of the animal. This will vary with the APL: the greater the APL the greater the ME required to provide a given amount of NE. Put the other way, the greater the APL the smaller the amount of NE an animal obtains from a given amount of ME. The NE method uses this fact.

The APL can be given a measure, **the APL factor**, which takes into account the size of the animal and the LWG, as in Table 22.

Using the APL factors, the ME value of feeds can be converted to NE values as in Table 23. Having determined the NE value of the feeds available it is then a relatively simple matter to calculate a ration to meet the animals' NE requirements.

By using Tables 20, 22, and 23 it is now possible to compile a ration from feeds of known analysis to provide sufficient energy for an animal of known liveweight to reach a specific LWG. The protein requirements must also be catered for and they are given below.

TABLE 22

Animal production levels (APL) of growing cattle

Liveweight	Liveweight gain (kg/day)							
	0.2	0.4	0.6	0.8	1.0	1.2	1.4	1.6
100	1.15	1.32	1.50	1.72	1.98	—	—	—
150	1.13	1.28	1.45	1.65	1.87	—	—	—
200	1.12	1.26	1.41	1.59	1.79	2.05	—	—
250	1.11	1.24	1.38	1.55	1.74	1.97	—	—
300	1.11	1.23	1.36	1.52	1.70	1.92	2.19	2.53
350	1.10	1.21	1.34	1.49	1.67	1.88	2.13	2.47
400	1.10	1.20	1.33	1.47	1.64	1.85	2.09	2.41
450	1.10	1.20	1.32	1.46	1.62	1.82	2.06	2.37
500	1.09	1.19	1.32	1.45	1.60	1.79	2.03	2.33
550	1.09	1.19	1.30	1.44	1.59	1.78	2.01	2.30
600	1.09	1.19	1.30	1.43	1.58	1.76	2.00	2.28

TABLE 23

Net energy (NE) values of feeds (MJ/kg DM) related to animal production level (APL)

APL	M/D of the feed												
	8.0	8.5	9.0	9.5	10.0	10.5	11.0	11.5	12.0	12.5	13.0	13.5	14.0
1.0	5.8	6.1	6.5	6.8	7.2	7.6	7.9	8.3	8.6	9.0	9.4	9.7	10.1
1.1	5.2	5.6	6.0	6.4	6.8	7.2	7.6	8.0	8.3	8.7	9.1	9.5	9.9
1.2	4.9	5.3	5.7	6.1	6.5	6.9	7.3	7.7	8.1	8.5	9.4	9.4	9.8
1.3	4.6	5.0	5.4	5.8	6.3	6 7	7.1	7.5	7.9	8.4	8.8	9.2	9.7
1.4	4.4	4.8	5.2	5.6	6.1	6.5	6.9	7.4	7.8	8.2	8.7	9.1	9.6
1.5	4.2	4.7	5.1	5.5	5.9	6.3	6.8	7.2	7.7	8.1	8.6	9.0	9.5
1.6	4.1	4.5	4.9	5.3	5.7	6.2	6.6	7.1	7.5	8.0	8.4	8.9	9.4
1.7	4.0	4.4	4.8	5.2	5.6	6.1	6.5	7.0	7.4	7.9	8.4	8.9	9.3
1.8	3.9	4.3	4.7	5.1	5.5	6.0	6.4	6.9	7.4	7.8	8.3	8.8	9.3
1.9	3.8	4.2	4.6	5.1	5.5	5.9	6.4	6.9	7.3	7.8	8.3	8.8	9.3
2.0	3.8	4.2	4.6	5.0	5.4	5.9	6.3	6.8	7.3	7.7	8.2	8.7	9.2
2.2	3.6	4.0	4.4	4.9	5.3	5.8	6.2	6.7	7.2	7.6	8.2	8.6	9.2
2.4	3.5	3.9	4.3	4.8	5.2	5.7	6.1	6.6	7.1	7.6	8.1	8.6	9.1
2.6	3.5	3.9	4.3	4.7	5.1	5.6	6.0	6.5	7.0	7.5	8.0	8.5	9.1

PROTEIN REQUIREMENTS

The protein requirements for a beef animal increase with the size of the animal and with the daily LWG. Table 24 combines the maintenance and production requirements.

TABLE 24

Total requirement of digestible crude protein (DCP) (g/day)

Liveweight (kg)	Liveweight gain (kg/day)						
	0.2	0.4	0.6	0.8	1.0	1.2	1.4
100	155	230	270	310	365	410	
150	215	295	335	380	435	485	540
200	250	330	380	425	480	535	590
250	270	355	410	460	520	570	625
300	295	385	435	490	545	595	650
350	310	395	450	510	570	615	675
400	325	410	465	520	575	625	685
450	340	415	475	540	595	650	710
500	365	440	505	565	615	675	735
550	375	450	515	575	625	700	745

COMPILING RATIONS FOR THE TABLES

Examples are given below of compiling rations for fattening cattle under some of the usual systems of production. The rations are compiled by using the net energy method and then checked against the metabolisable energy table (Table 20).

Example 1. Finishing cattle on a silage/cereal diet

Many cattle on an 18-month beef system are finished on a silage/cereal diet. The animals come in off the grass at around 350 kg liveweight in early October and go on to a ration of, say, 2 kg barley and silage to appetite. The target for LWG is about 0.8 kg/day. The DMI of a 350-kg animal is 8.0 kg/day (Table 17).

The analysis of the feeds is assumed to be:

Silage: 9.0 ME, 100 DCP, 200 DM.
Barley: 13.0 ME,   82 DCP, 850 DM.

Step 1. Determine the **NE requirements** (Table 21):

350-kg animal: LWG 0.8 kg = 42.6 NE.

Step 2. Determine the **APL factor** (Table 22):

> APL factor for this animal: 1.49 (round to 1.5).

Step 3. Determine the **NE value of the feeds** (Table 23):

> At an APL factor of 1.5 the feeds have an NE value:
> Barley (13.0 ME) = 8.6 NE.
> Silage (9.0 ME) = 5.1 NE.

Step 4. Calculate the **NE contribution** of the **set part of the ration**:

> 2.0 kg barley as fed = 1.7 kg DM barley $\times$ 8.6 = 14.6 NE.

Step 5. Calculate how much **silage** will be needed **to make up the balance;**

> Balance of NE required: 42.6 − 14.6 = 28.0 NE.
> Silage required: $\dfrac{28.0}{5.1}$ = 5.5 kg DM silage = 27.5 kg as fed.

Step 6. Check the quantities are **within the DMI**:

> 1.7 kg barley + 5.5 kg silage = 7.2 kg DM, i.e. within the DMI of 8.0.

It will be seen below that the M/D of this ration is only 10.0 and at this level the DMI would probably be nearer 7.0 than 8.0.

Step 7. Check the amount of **protein supplied by the ration** and add a protein supplement to the barley if necessary:

> The DCP requirement for this animal is 510 g (Table 24).

> At 100 g DCP/kg DM the silage alone will meet this requirement. There would be no need to add a protein supplement to the barley.

The **ration** can be **checked through the ME system** (Table 25).

TABLE 25

Feed as fed	kg DM		ME/kg DM		ME in the ration
2.0 kg barley	1.7	$\times$	13.0	=	22
27.5 kg silage	5.5	$\times$	9.0	=	50
	7.2				72

An M/D of 10.0.

Table 20 shows that for a 350-kg animal, on a ration with an M/D of 10.0 and supplying 72 ME, the LWG would be 0.8 kg/day.

Example 2. Final stages of fattening on a silage/cereal diet

In 3 or 4 months the animal will reach 450 kg liveweight and be in the final stages of fattening. It is usual to increase the barley ration over this period up to 5.0 kg or more. The ration can be calculated as before, first by the NE system then checked by the ME system.

450-kg animal with a target LWG of 1.0 kg/day.

The animal: 56.9 NE   The feeds: as before except that
            1.62 (say 1.6) APL   the NE values are
            9.0 DMI   lower due to the
            595 DCP   increased APL:
                                        Barley 8.4 NE.
                                        Silage 4.9 NE.

5.0 kg barley as fed provides 4.25 kg DM $\times$ 8.4 $=$ 35.7 NE.
Balance to be supplied by the silage $=$ 21.1 NE,
supplied by $\dfrac{4.33}{21.2} = 4.9$ kg DM silage $=$ 22.0 kg as fed.

The ration can now be checked for total DM, NE, ME, DCP and M/D (Table 26).

TABLE 26

Feed as fed	kg DM		Analysis of DM		NE	ME	DCP
5.0 kg barley	4.25	$\times$	8.4	$=$	35.7		
		$\times$	13.0			$=$ 55	
		$\times$	82				$=$ 349
22.0 kg silage	4.33	$\times$	4.9	$=$	21.2		
		$\times$	9.0			$=$ 39	
		$\times$	100				$=$ 433
	8.58				56.9	94	782

An M/D of 10.9.

Table 20 predicts a LWG of 1.0 kg/day for an intake of 94 ME at an M/D of 10.9. The 8.6 kg DM is within the DMI of 9.0.

Again the concentrate need only consist of barley as the protein needs are covered without the use of protein supplement. It should be noted that the protein content of the silage in the example is quite good, and many silages may not be as high as this. If there was any doubt about the protein content of the bulk feed, whatever it was, it would be sound practice to add 10–15% of a protein supplement to the barley. Proprietary supplements are usually reinforced with minerals and vitamins.

Both these rations depend on the animals being able to fill themselves to capacity on the silage. If for some reason they are not able to do so, then the predicted growth rates will not be attained. It can be seen that it is the quantity of barley that has been increased while the amount of silage is nearly the same.

This is a feature of this fattening system: as the winter goes on and the animals increase in size, their increased requirements are met by feeding more concentrate while the intake of silage remains fairly constant. This has the added effect of increasing the M/D of the ration which again helps to maintain feed intake and thereby the LWG of the stock.

BARLEY BEEF PRODUCTION

This is the most intensive form of beef production where young cattle are fed *ad lib* on a concentrate diet normally based on barley. The proportions in the mix are usually 85% barley to 15% protein–mineral–vitamin compound. The analysis of the compound may only be known to the manufacturer but it can be assumed to be similar to that of groundnut cake. The analysis of the mix is given in Table 27.

TABLE 27

Feed as fed	kg DM		Analysis of DM		ME		DCP
8.5 kg barley	7.3	×	13.0	=	95		
		×	82			=	599
1.5 kg supplement	1.3	×	12.9	=	17		
	—	×	449			=	584
	8.6				112		1183

The diet has an M/D of $\dfrac{112}{8.6}$ = 13.0 and a DCP level of 137.

It has been shown that stock, whose basic diet is concentrate, perform better when a small amount of long fodder is included in the daily ration. This not only assists in the digestion process but also helps to maintain feed intake. The long fodder may be in the form of hay or barley straw. When cattle are regularly bedded with barley straw, they are assumed to consume a certain amount of it. Example 3 assumes an intake of 0.5 kg DM in the form of long fodder in the early stages rising to 1.0 kg towards the end of the fattening period. The DMI in the form of concentrate is adjusted accordingly. As the analysis of the long fodder is not known and can be variable, no allowance has been made of any small contribution it might make to the ration overall. The M/D of the ration will, however, be a little lower than for a ration of concentrate alone and it has been assumed that the M/D will be about 12.5 rather than 13.0. The standard DMI has been increased in the example below by 0.5 because of the high M/D.

Example 3. Barley beef production

This example is given in tabular form (Table 28) taking the 150-kg reared calf of around 20 weeks of age through to the finished beast of 350–400 kg liveweight. Throughout the period the animal is fed on an *ad lib* diet of the barley mix given above with a small allowance of long fodder. The predicted LWG is based on the ME allowances in Table 20.

Under such a feeding regime it can be seen that the consumption of the mix would reach nearly 9.0 kg as fed, giving a LWG of 1.3 kg with the cattle finishing at from 10 to 12 months of age. Such figures are common for this type of production.

A TRADITIONAL FATTENING RATION

There are still many areas where store cattle are finished in the traditional way using arable crops such as seeds hay, straw, roots, and barley. An example of a typical ration is given (Table 29) for a 400-kg beast required to fatten to 1.0–1.2 kg/day.

TABLE 28

Barley beef production: 150 kg liveweight to slaughter

Live-weight	DMI (kg DM)	DMI from long fodder	DMI of barley	Intake of nutrients form barley mix ME	DCP	Predicted LWG (kg/day)	Days to gain 50 kg	Days to reach this live-weight
150	4.5	0.5	4.25	50	413	1.0	50	140
200	5.5	0.5	5.00	67	685	1.2	42	190
250	6.5	0.75	5.75	72	787	1.25	40	232
300	7.5	0.75	6.75	84	925	1.25	40	272
350	8.5	1.0	7.5	94	1028	1.3	38	312
400	9.0	1.0	8.0	100	1096	1.3	38	350

Example 4. A traditional fattening ration

TABLE 29

Feed as fed	kg DM		Analysis of DM		ME	DCP
3.5 kg hay (medium quality)	3.0	×	8.4	=	25	
		×	39.0			= 117
25.0 kg swedes	3.0	×	12.8	=	38	
		×	91.6			= 273
3.0 kg barley	2.6	×	13.0	=	34	
		×	82.0			= 213
	8.6				97	= 603

$$M/D = 11.3$$

At an M/D of 11.3 a ration containing 97 ME should fatten a 400-kg animal at 1.1 kg/day. The protein content is only just adequate, and it would be advisable to add 10–15% of a high protein supplement. Traditionally this would have been groundnut cake (or even cotton cake). The addition of the protein would have little affect on the ME of the mix but would give a more satisfactory level of protein in the ration as a whole.

MINERALS AND VITAMINS

The needs of fattening cattle for minerals are not as great as for dairy cows. Nevertheless, they have a definite requirement, and at the higher levels of production a shortage of minerals could be a limiting factor. Calcium and phosphorus are the two which could be in short supply, and it is usual to include these in the concentrate mix as a mineral supplement.

It is also advisable to add vitamins A, D and E in the form of a proprietary supplement to the concentrate mixes of housed cattle on winter rations. Although silage and hay supply considerable quantities of vitamin A and vitamin E, only sun-dried hay contains any quantity of vitamin D. Roots such as swedes and barley contain none of these. A vitamin supplement should be a feature of the diets of all growing cattle under these conditions.

QUESTIONS—BEEF RATIONING

1. What general conditions make the rationing of beef cattle somewhat different from rationing dairy cows?
2. How does the DMI of a beef animal vary with its liveweight?
3. What two factors affect the efficiency of utilisation of energy in growing cattle?
4. State the relationship between each of the two factors and the efficiency with which the dietary energy is utilised.
5. What is the relationship between ME and NE? (see also Chapter 10).
6. What is an APL factor?
7. Predict the daily LWG for the following animals on the rations given:

	Liveweight (kg)	M/D of the ration	ME in the ration	Predicted LWG
(a)	200	12	47	
(b)	250	11	55	
(c)	350	10	73	
(d)	400	13	88	
(e)	450	12.5	85	

8. For the following animals growing at the rate shown, give the NE allowances and the APL factor.

	Liveweight	LWG	NE allowance	APL
(a)	200	0.4		
(b)	300	0.8		
(c)	400	1.0		
(d)	450	0.6		
(e)	450	1.2		

9. For each of the animals in question 8 what would be the NE value of the following feeds?

	Feed	ME
(a)	Hay	9.0
(b)	Silage	10.2
(c)	Barley	13.7
(d)	Potatoes	12.5
(e)	Grazing	10.5

CHAPTER 13

Rations for sheep

SHEEP are kept under a wide range of conditions, and from a rationing point of view it is not easy to make generalisations. On some farms sheep are the main enterprise, but more often than not they are then kept under extensive conditions on hill and upland areas. For much of the year they rely on what the hill has to offer for their nutrition, and, indeed, they are the main class of stock for utilising this less productive land.

On other farms the sheep are a supplementary enterprise and are there to utilise crop material and surplus grazing which would otherwise be wasted. In these circumstances it may appear that the sheep is again left to ration itself as its main purpose is to make something out of nothing.

At the other extreme it is possible to house lambs and feed them intensively, and their nutrition is then completely in the hands of the feeder. Ewes also may be housed in the later stages of pregnancy, and, again, it is necessary to be able to ration them accurately.

Whether housed, folded, or extensively grazed, the nutrition of sheep is, nevertheless, constantly monitored in order to make sure they are in the condition the system demands while making most efficient use of the feed available. There are, however, certain periods in the production cycle when the nutrition is critical, and most of the obvious rationing of sheep revolves round these periods.

For the ewe there are three periods: at **tupping**, in **late pregnancy**, and in the **early part** of the **lactation**. In fat-lamb production the **final fattening period** is a time when the sheep are rationed carefully. This ensures they grow satisfactorily in relation to the bulk feeds available and the dates it is intended to market them.

FEEDING STANDARDS

As with other stock, theoretical feeding requirements are available for every class and condition of sheep, but feeding sheep is still a long way from being an exact science. At the best the tables can only be an approximate guide, and any ration based on them should be monitored carefully to prove it in practice. No sheep is fed individually, and the standards can only apply to the group as a whole. Grazing nearly always forms the base of the diet, and the rationing that is practised is in relation to it. It is impossible to measure the amount of herbage sheep are consuming and thereby the amount of supplementary feed to make up the balance. Sometimes sheep decide for themselves, and as soon as the grazing is plentiful leave the hand-fed supplements untouched. Generally speaking, therefore, the **only reliable guide is the condition and behaviour of the sheep.**

TABLE 30

Daily allowances for DMI, ME, and DCP of ewes at different points in the breeding cycle

Liveweight (kg)	Appetite (DMI kg*)	Maintenance		Tupping		At lambing		Early lactation	
		ME	DCP	ME	DCP	ME	DCP	ME	DCP
30	0.77	4.8	39	8.1	50	S 7.7	78	S 15.3	152
						T 9.2	89	T 22.3	210
60	1.53	7.8	53	14.4	64	S 11.9	99	S 20.3	211
						T 14.7	113	T 27.6	278
90	2.30	10.8	67	20.6	79	S 16.2	120	S 22.3	225
						T 20.1	137	T 29.6	292

*Reduce by 15% in late pregnancy and increase by 40% at peak lactation.

FEEDS AND PALATABILITY

Palatability in the sheep diet is of utmost importance. Sheep are fastidious feeders and would rather starve than take unpalatable feed. Of the protein feeds, decorticated groundnut and linseed cake are the traditional feeds, but nowadays, as with other stock, compound protein–mineral–vitamin supplements are often used. Of the energy feeds, flaked maize, kibbled locust beans, and molassed sugar-beet pulp are all very palatable. Rolled barley, or in some areas rolled oats, usually form the basis of the concentrate mix.

Hay must also be of good quality, especially for suckling ewes, and musty hay is completely unacceptable to sheep.

RATIONS FOR EWES AT THE CRITICAL POINTS IN THE BREEDING CYCLE

Feeding at tupping time

Lambs are normally weaned by mid-June or July, and from then until the autumn the ewes are fed as cheaply as possible. As tupping time approaches the aim is to have them improving in condition in the belief that they will then tend to shed two, or more ovules at ovulation and there will be a greater chance of producing twins. The recommended feeding standards, in the period before and during the time the tups are with the ewes, are considerably above the normal maintenance requirements. For a 60-kg ewe the comparative figures are (from Table 30):

Maintenance: 7.8 ME   53 DCP
At tupping: 14.4 ME   64 DCP

To achieve this increase it is usual to bring the ewes on to better pasture or green crop such as rape grown for the purpose, about 2 weeks before the tups are put in. It is then important to see that the higher level of nutrition is maintained throughout the period which may be around 8 weeks. In this time the weather may deteriorate with the grass decreasing in quantity and quality, and supplementary feeding then becomes necessary.

A concentrate mix of barley and a protein supplement similar to that fed in late pregnancy (85% barley, 15% protein feed) should be fed in good time to avoid any setback. The daily rate may vary from 0.2 to 0.5 kg as fed. Hay might also be on offer if the state of the grazing warrants it.

Last 6 weeks of pregnancy

It is common practice to start feeding ewes with a supplementary concentrate feed some 6 weeks before lambing at, say, 0.25 kg/day rising to nearly 1.0 kg/day, depending on the size of the ewe and the quality of the grazing.

For a 60-kg ewe the ME requirement per day rises from around 8 ME to 12–15 ME (Table 30) (single lamb, 11.9 ME; twins, 14.7 ME). Taking the ME figures for a ewe carrying twins, the ME requirements in the last weeks of pregnancy are as Table 31 shows.

TABLE 31

Empty ewe	Weeks before lambing				
	8	6	4	2	Birth
7.8	8.1	9.4	10.9	12.7	14.7

A 60-kg ewe has a DMI of 1.53, and if the ewe's diet is made up from two feeds, say,

Hay:	8.4 ME	39 DCP,
Barley mix:	13.5 ME	137 DCP,

—from an energy point of view the hay alone would meet her requirements to within 3 weeks of lambing. In the last 3 weeks, however, the situation changes quite dramatically when an increased nutrient requirement coincides with a depression in appetite. At birth the M/D of the ration needs to be

$$\frac{14.7 \text{ ME}}{1.3 \text{ DMI}} = 11.3,$$

—which would require a diet of nearly 60% barley mix and 40% hay. The ration would then be as Table 32 shows.

TABLE 32

Feed as fed	kg/DM		ME in feed		ME in ration
0.6 kg hay	0.52	×	8.4	=	4.4
0.9 kg barley mix	0.78	×	13.5	=	10.5
	1.3				14.9

In most circumstances it would not be wise to leave the feeding of concentrates until the last 3 weeks of pregnancy. Only with housed sheep can an adequate supply of bulk feed be guaranteed. Outside sheep may be relying on grazing alone, and by January this can be in short supply. The increasing needs of the sheep are likely to outstrip the nutrition available from the grazing, and it becomes necessary to meet more of the needs from hand feed—hay and concentrates.

The protein content of the ration must not be forgotten, and grazing or hay alone may not meet the protein needs of the ewe in late pregnancy. A 60-kg ewe needs 53 DCP when empty rising to 113 DCP by the time it lambs.

Table 33 shows the extent to which a concentrate feed (85% barley, 15% high protein feed) contributes to the nutrition of a 60-kg ewe carrying twins as she approaches lambing: fed from 0.2 to 1.0 kg per day over the last 8 weeks.

TABLE 33

		Weeks before lambing								
		8		6		4		2		Lambing
		ME	DCP	ME	DCP	ME	DCP	ME	DCP	ME	DCP
Feeding standards		8.1	60	9.4	70	10.9	82	12.7	95	14.7	113
Con-	as fed	0.2		0.25		0.6		0.8		1.0	
cen-	kg DM	0.17		0.21		0.51		0.68		0.85	
trate:	ME, DCP	2.2	24	2.7	29	6.6	68	8.8	93	11.0	116
Bal-											
ance:	ME, DCP	5.9	36	6.7	41	4.3	14	3.9	2	3.7	—
Hay*	as fed	1.0		1.0		0.7		0.6		0.5	
	kg DM	0.9		0.9		0.6		0.5		0.4	
	ME, DCP	7.1	33	7.1	33	5.0	24	4.2	19	3.6	16
Complete											
ration:	ME, DCP	9.3	57	9.8	62	11.6	92	13.0	112	14.6	132

*Hay: 8.4 ME, 39 DCP.

It can be seen that early on in the feeding period the greater part of the nutrients are supplied by the bulk feed, but as lambing approaches the concentrate plays the major role and less reliance is placed on the hay.

All above rations are well within the standard DMI of 1.53 for this size of ewe. In the last 2 or 3 weeks of pregnancy appetite is reduced by nearly 20% but even so the final ration shown is still within the appetite of the ewe.

Sheep are, of course, ruminants, and are able to synthesise their own protein from a source of nitrogen providing they have an adequate supply of energy, minerals, vitamins and a roughage feed. In-lamb ewes may be housed in late winter and fed a ration of barley, 0.75–1 kg/day and **urea** in the form of **molasses based liquid feed.** Barley **straw** should be available ad lib and also a supply of **minerals.** Such a system is particularly suitable to a farm where the sheep are a complementary enterprise to a dairy herd and should not compete for hay or silage in winter.

Early lactation

The principles of feeding during lactation are the same for the ewe as for the cow. There is a greater nutritional need as lactation gets under way and, after an initial depression, an increase in appetite. It is nevertheless impossible for a ewe to maintain a high level of milk production without loss in bodyweight and it is essential therefore that she lambs down in good condition. Concentrate feeding should continue during the suckling period for 4–6 weeks or until such times as the lambs are taking creep feed and grazing is plentiful for both ewes and lambs.

A 60-kg ewe with twins needs 23.3 ME and 278 DCP per day in the first month of lactation. It would take 2 kg of the barley mix quoted above (13.0 ME, 137 DCP) and with an appetite of only 1.53 DMI the ewe could not cope with this amount of feed. Fortunately the appetite increases by up to 40% and, for this ewe, a DMI of up to 2.1 kg can be expected. To take in 23.3 ME within 2.1 kg DM the ration must have an M/D of 11.1.

Using the same hay and barley mix as fed in late pregnancy, a ration of 40% DM hay and 60% DM barley mix will be needed to give an M/D of 11.1.

The daily ration would be:

0.85 kg DM hay	×	8.4	=	7.1
1.25 kg DM mix	×	13.0	=	16.2
2.1 kg DM				23.3 ME

This ration is low in protein—the hay supplies 33 DCP leaving 245 DCP to be made up by the barley mix. The protein level of the mix needs to be:

$$\frac{245}{1.25} = 196$$

A mix of approximately 60% barley and 30% high protein supplement (e.g. groundnut) has this level of protein (see protein balancing chart, Fig. 3):

0.7 kg DM barley	×	82	=	57
0.3 kg DM groundnut	×	449	=	135
1.0 kg DM mix				192 g DCP

1.25 kg DM mix (1.5 kg as fed) supplies 240 g DCP.

The inclusion of this amount of groundnut (12.9 ME) does not materially alter the ME value of the mix.

If the **full nutritional requirements** of the lactating ewe are to be met from a diet of hay and barley mix, then the ration shown in Table 34 is required.

TABLE 34

Feed as fed	kg DM		Analysis of DM		ME	DCP
1.0 kg hay	0.85	×	8.4	=	7.1	
		×	39	=		33
1.5 kg barley mix	1.25	×	13.0	=	16.2	
		×	192	=		240
	2.1				23.3	273

In practice it would not be usual to feed more than 1.0 kg of concentrate even to large ewes nor to include more than 25% high protein feed in the mix. There are three factors that have brought this about in this example:

(a) the number of lambs per ewe;
(b) the ability of the ewe to milk off her back;
(c) the quality of the bulk feed.

The example has been based on a ewe with twins whereas the flock as a whole will contain both twins and singles and on average the requirement will be lower. Furthermore, if the ewe is in good condition she will be expected to contribute about 2 MJ of energy per day, which would represent a loss of 0.5 kg liveweight per week. The third factor is the bulk feed—hay of only moderate quality has been used in the example. Good hay could have 20% more energy value. The feed of real value, however, is spring pasture grass which can have 50% more energy value and 200% more protein than moderate quality hay, and there is of course, nothing better to bring the ewes into milk. Obviously as soon as the grass becomes available the protein level of the concentrate mix can be reduced and also the overall level of concentrate feeding.

From an economic standpoint, when lambs are being sold fat off the ewes, it may be better to spend money on creep feed for the lambs rather than overdo the concentrates for the ewes.

Minerals for lactating ewes

A 60-kg ewe with twins requires:

Ca, 16; P, 9.9; Mg, 3.0; Na, 2.1.

The ration of hay and concentrate quoted for early lactation would only meet one-third of the calcium requirements and would also be deficient in phosphorus.

A daily supplement of 30 g of steamed bone flour would make up the deficiencies. A 2–3% inclusion rate in the concentrate mix would be satisfactory. An alternative would be to have an ad lib supply of a proprietary mineral mix always available, but in one form or another a mineral supplement is essential for lactating ewes.

TABLE 35. *Net energy allowances for growing sheep*

Liveweight (kg)	Liveweight gain (g/day)					
	50	100	150	200	250	300
20	3.8	4.6	5.6	6.5	7.5	8.4
25	4.4	4.9	6.2	7.2	8.2	9.3
30	4.9	5.8	6.8	7.8	8.9	10.0
35	5.5	6.4	7.5	8.6	9.7	10.8
40	6.0	7.0	8.1	9.2	10.4	11.6
45	6.6	7.6	8.8	10.0	11.2	12.4

TABLE 36. *Animal production levels (APL) of growing sheep*

Liveweight (kg)	Liveweight gain (g/day)					
	50	100	150	200	250	300
20	1.38	1.60	1.94	2.29	2.65	3.02
25	1.25	1.53	1.84	2.15	2.48	2.81
30	1.23	1.49	1.77	1.05	2.35	2.65
35	1.21	1.45	1.71	1.98	2.25	2.54
40	1.20	1.43	1.67	1.92	2.18	2.44
45	1.19	1.41	1.64	1.88	2.12	2.37

TABLE 37. *Net energy values of feeds for growing sheep (MJ/kg DM)*

APL	M/D of the feed						
	8	9	10	11	12	13	14
1.0	5.6	6.3	7.0	7.7	8.4	9.1	9.8
1.1	5.1	5.9	6.6	7.4	8.1	8.9	9.7
1.2	4.8	5.6	6.3	7.1	7.9	8.7	9.6
1.3	4.5	5.3	6.1	7.0	7.8	8.6	9.5
1.4	4.3	5.1	6.0	6.9	7.6	8.5	9.3
1.5	4.2	5.0	5.8	6.7	7.5	8.4	9.3
1.6	4.1	4.9	5.7	6.6	7.4	8.3	9.3
1.7	4.0	4.8	5.6	6.5	7.4	8.3	9.2
1.8	3.9	4.7	5.5	6.4	7.3	8.2	9.2
1.9	3.8	4.6	5.4	6.3	7.2	8.2	9.1
2.0	3.7	4.5	5.4	6.2	7.2	8.1	9.1
2.2	3.6	4.4	5.3	6.1	7.1	8.0	9.1
2.4	3.5	4.3	5.2	6.1	7.0	8.0	9.0
2.6	3.4	4.2	5.1	6.0	6.9	7.9	9.0
2.8	3.4	4.2	5.0	5.9	6.9	7.9	8.9
3.0	3.3	4.1	5.0	5.9	6.8	7.8	8.9

TABLE 38. *Daily ME allowances for growing sheep (MJ/day)*

Weight (kg)	DMI (kg)	M/D	Daily LWG (g)					
			50	100	150	200	250	300
		9	6.3	—	—	—	—	—
		10	6.1	8.0	—	—	—	—
		11	5.9	7.7	—	—	—	—
20	0.8	12	5.8	7.4	9.2	—	—	—
		13	5.7	7.2	8.8	10.5	—	—
		14	5.6	7.0	8.5	10.0	—	—
		9	7.1	9.4	—	—	—	—
		10	6.9	9.0	—	—	—	—
		11	6.8	8.6	10.6	—	—	—
25	0.9	12	6.6	8.3	10.1	—	—	—
		13	6.5	8.1	9.8	11.5	—	—
		14	6.4	7.9	9.4	11.0	12.7	—
		9	8.0	10.4	—	—	—	—
		10	7.8	9.9	—	—	—	—
		11	7.6	9.5	11.6	—	—	—
30	1.05	12	7.4	9.2	11.1	13.1	—	—
		13	7.3	9.0	10.7	12.5	—	—
		14	7.2	8.8	10.4	12.1	13.8	—
		9	8.8	—	—	—	—	—
		10	8.6	—	—	—	—	—
		11	8.4	10.5	12.7	—	—	—
35	1.15	12	8.3	10.1	12.1	14.2	—	—
		13	8.1	9.9	11.7	13.6	15.6	—
		14	8.0	9.6	11.3	13.1	14.9	—
		9	9.7	—	—	—	—	—
		10	9.4	11.8	—	—	—	—
		11	9.2	11.4	—	—	—	—
40	1.30	12	9.1	11.1	13.1	—	—	—
		13	9.0	10.8	12.7	14.7	—	—
		14	8.9	10.5	12.3	14.2	16.1	—
		9	10.5	—	—	—	—	—
		10	10.3	12.7	—	—	—	—
		11	10.1	12.3	14.7	—	—	—
45	1.45	12	9.9	12.0	14.2	—	—	—
		13	9.8	11.7	13.7	15.8	—	—
		14	9.7	11.4	13.3	15.2	17.2	—
		9	10.9	—	—	—	—	—
		10	11.1	13.7	—	—	—	—
		11	10.9	13.3	15.8	—	—	—
50	1.41	12	10.8	12.9	15.2	17.6	—	—
		13	10.7	12.6	14.7	16.9	19.2	—
		14	10.5	12.3	14.3	16.3	18.4	20.6

TABLE 39

Protein allowances for growing sheep (g DCP/day)

Liveweight (kg)	Daily LWG (g/day)					
	50	100	150	200	250	300
20	45	57	69	81	93	105
25	46	58	70	82	94	106
30	48	60	72	84	96	108
35	51	63	75	87	99	111
40	53	65	77	89	101	113
45	55	67	79	91	103	115
50	57	69	81	93	105	117

FINISHING HOGGS

Lambs which are not fattened off the grass may be stored over the late summer and finished off either on grass or roots in the autumn and winter. Hoggs from hill breeds may only reach 30–35 kg liveweight whereas hoggs from heavier breeds may reach 50 kg or more.

The principles of rationing growing sheep are the same as for growing cattle. As the size of the sheep increases, so does the energy allowance for maintenance, and the higher the daily LWG and the APL, the greater the energy allowance for growth. The efficiency with which the animal uses the energy in its diet again varies with the M/D; the higher the M/D the more efficiently the energy is used.

One problem with sheep is the question of appetite. The standard DMI allowances vary from one authority to another just as the appetite of any given size of sheep varies with the type of feed in its diet. Palatability plays a part but so does the nature of the feeds. Sheep on chopped swedes, for example, are known to consume more dry matter per day than sheep on folded swedes and more again when the same-sized sheep are on a ration of silage and barley. Clearly the overall M/D has an influence on DMI. The condition of the sheep may also be a factor, and the loss of teeth in hoggs is notorious for causing a reduced DMI and a consequent loss of weight. Of all the figures given in the tables, those for DMI should be treated with the most caution. Having devised a ration based on an assumed DMI, if it is at all feasible, some form of check should be carried out on the weights of feed actually being taken. If this is not possible, then the performance of the sheep should be carefully monitored.

As for beef cattle, there are the two methods of calculating rations for growing sheep—the **net energy method** and the **metabolisable energy method**. The NE method is the best for actually devising the ration, whereas the ME method is the easiest for predicting the likely performance from a given ration. In the examples that follow this procedure is followed, and one method is tested against the other.

The three tables needed for the NE method are Tables 35–37. Table 38 gives the ME allowances for growing sheep and Table 39 the protein requirements.

Example 1

A ration for a 30-kg sheep required to grow at 100 g/day, folded on kale with barley as a supplement if needed

The sheep:	NE energy,	5.8	(from Table 35)
	APL factor	1.49	(from Table 36)
	DMI	1.05	(from Table 38)
	Protein (DCP)	60	(from Table 39)

The feed:	Kale:	140 DM	11.0 ME	6.7 NE	(from Table 37)
	Barley:	860 DM	13.7 ME	8.4 NE	(from Table 37)

To take in 5.8 NE the animal would only need to consume $\frac{5.8}{6.7}$ kg — 0.86 kg DM kale, which is well within its appetite, and there would be no need to feed any barley. There is no reason, in fact, why the sheep should not take 1.05 kg DM kale giving an intake of 11.55 ME at an M/D of 11. Table 38 predicts that the LWG would be nearly 150 g/day, which would be quite usual for this class of sheep folded on a green crop.

Example 2

It might be necessary to increase the LWG of the same 30-kg sheep to 200 g/day using the same feeds. This is a high level of production which is reflected in the APL factor and the amount of concentrate needed to achieve it.

The sheep:	7.8 NE	The feeds:	as before except that the
	2.05 APL		NE value is lower because
	1.05 DMI		of the ·increased APL:
	84 DCP		Barley 8.1 NE.
			Kale 6.2 NE.

The combination of kale and barley to provide 7.8 NE can be determined by using the **feed combination chart** (Fig. 4). Assuming a DMI of

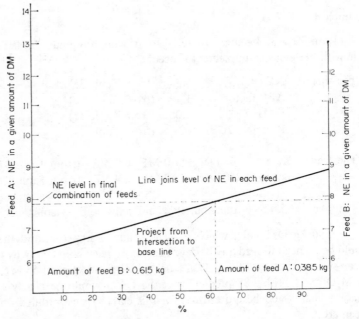

FIG. 4. FEED COMBINATION CHART. To determine the proportions of two feeds that will supply a required amount of NE.

From Example 2 on p. 124/130:
   Kale (feed A) provides 6.2 NE in 1.0 kg DM,
   Barley (feed B) provides 8.8 NE in 1.0 kg DM.

Level of NE required in 1.0 kg DM of the combined feeds, 7.8.

1.0 kg (actual 1.05), if the ration was entirely kale the sheep would consume 6.2 NE; if it was all barley it would take in 8.1 NE. The combination of the two feeds shown by the chart to give 7.8 NE is:

Kale	0.15 kg DM	×	6.2	=	0.9 NE		
		×	11.0			=	1.7 ME
Barley	0.85 kg DM	×	8.1	=	6.9 NE		
		×	13.0			=	11.0 ME
	1.0 kg DM				7.8 NE		12.7 ME

$$M/D = 12.7$$

Table 38 predicts that the LWG for this sheep, with an intake of 12.7 ME in a ration with an M/D of 12.7 will be 200 g. The feeds as fed would be 1.0 kg barley and just over 1.0 kg kale. In practice, because of the high M/D and high palatability of the feeds, the DMI could be above the 1.0 kg suggested. For example, a ration of 0.6 kg barley and 4.25 kg kale (as fed) would provide the same energy intake but in 1.1 kg DM. This would probably be a cheaper ration.

Either ration would meet the protein requirements without difficulty.

Example 3

A ration for a 45-kg hogg fattening at 200 g/day, folded on turnips, with an allowance of 0.25 kg DM hay (0.3 kg as fed) and a cereal mix of 50 :50 barley and oats.

The sheep:		The feeds:	DM	ME	DCP	NE
10.0	NE	Hay	850	8.5	35	4.2
1.88	APL	Turnips	90	11.2	73	6.5
1.45	DMI	Cereals	860	12.5	77	7.7
89	DCP					

The 0.25 kg DM hay provides 1.05 NE leaving 8.95 NE to be supplied by the turnips and cereal mix in 1.2 kg DM.

The combination of turnips and cereal can be determined from the feed combination chart using the following NE figures for the two feeds:

$$1.2 \text{ kg DM turnips} \times 6.5 = 7.8 \text{ NE}$$
$$1.2 \text{ kg DM cereals} \times 7.7 = 9.2 \text{ NE}$$

The chart indicates:

$$25\% \text{ turnips } 0.25 \times 1.2 = 0.33 \text{ kg DM turnips}$$
$$75\% \text{ cereal } 0.75 \times 1.2 = 0.9 \text{ kg DM cereals}$$

The complete ration is shown in Table 40.

TABLE 40

Feed as fed	kg DM		Analysis of DM		NE		DCP		ME
0.3 kg hay	0.25	×	4.2	=	1.05				
		×	35			=	9		
		×	8.5					=	2.1
3.3 kg turnips	0.3	×	6.5	=	2.0				
		×	73			=	22		
		×	11.2					=	3.4
1.0 kg cereal	0.9	×	7.7	=	6.9				
		×	82			=	74		
		×	12.5					=	11.3
	1.45				9.95		105		16.8

$$M/D = \frac{16.8}{1.45} = 11.6.$$

Table 38 predicts a LWG of 200 g/day for an intake of 16.4 ME at an M/D of 12.0.

The DCP level is satisfactory. Had it not been, a high protein feed such as groundnut cake (12.9 ME, 449 DCP) could have been added to the cereal mix without altering the ME or NE value.

The mineral requirements of this sheep are:

	Ca	P	Mg	Na
	8.0g	3.9g	0.88g	1.1g
and the ration supplies	3.0g	5.0g	1.8g	1.3g

The only mineral in short supply is calcium, and root crops are notoriously short of this element. It would be necessary, therefore, to add a calcium supplement to this ration. Had the diet contained a greater proportion of green fodder or hay, the mineral balance could have been better, but with less concentrate the phosphorus may have become critical.

It is obvious that a small amount of mineral supplementation is desirable and this can be added to the concentrate mix in the usual way. Where sheep are not receiving a concentrate feed (as in Example 1), then a mineral supply should be available in an *ad lib* feeder.

As with beef cattle, the main vitamins are A, D, and E. If the hay in the ration was sun-dried hay of reasonable quality, then it could meet the vitamin A and D requirements and some of the vitamin E. The cereals could make up the balance of vitamin E. The turnips would not contribute anything to the vitamin content. It is debatable, therefore, if vitamin supplementation is necessary, but, once again, it is a simple and cheap matter to add a supplement to the concentrate mix.

FEEDING SHEEP IN GENERAL

As with other stock that has unlimited access to bulk feed—either grazing or conserved fodder, a daily ration is in two parts, i.e. the concentrate, which is normally trough fed, and the *ad lib* bulk feed. Only the concentrate allowance is controlled by the feeder, and he should make every effort to see that each animal receives the calculated quantity. The major problem is to make sure the animal then fills up to capacity on the bulk feed as the proper rationing of the animal depends on this happening. This is difficult enough under a more or less controlled environment of a cattle or sheep shed with a given supply of feed of known quality. It is much more difficult under grazing conditions, where the quantity and quality of the feed can be so variable and the weather can play havoc with feed intake.

Under these circumstances the skill of the stockman is paramount in recognising any sign which suggests that the stock are not taking the feed that is expected of them. He then needs to decide if the conditions are likely to be only temporary or more permanent and what adjustments are necessary to the part of the ration he does control. This is the very essence of stock feeding for which no feeding table or book on rationing can be a substitute.

HILL SHEEP

Vast areas of mountain and hill land in this country are used for sheep farming, and it is through these animals that such land is made productive. Hill farms normally have an area of better land (in-bye) adjacent to the farmhouse and an area of hill which may be common land or enclosed. When the hill is open common the rate of stocking tends to be higher than on comparable enclosed hill. The main reason for this is that a certain number of ewes are needed to maintain an individual's right to a certain area of hill and if the number is reduced, sheep from neighbouring flocks encroach on the area (heaft/hirsel) in search of better grazing. Under this system, which is typical of many dales areas in north-west England, a considerable reliance has to be placed on the in-bye land to supplement the fell grazing. When the hill is enclosed the stocking is often appreciably less and as a result the land carries ewes for a longer period of the year—even the whole year round.

There is usually a great difference in the quality of the hill grazing compared with that of the better land and the hill ewe is subjected, therefore, to different levels of nutrition at different times of the year— particularly so under the common-land system where the ewe spends about 8 months of the year at the hill and 4 months on the in-bye. It is vitally important that the transition from one type of grazing to the other is made gradually. Most hill farms have a number of fields or allotments of different quality grazing through which the sheep pass, over a period of days, when being brought down or on returning to the hill.

The in-bye land acts as a supplement to the hill both as supplementary grazing and in the production of hay for feeding in winter. Good hay must always be the main standby for winter feed on the hill farm, and the success or failure of the hay crop largely determines how well—and how profitably—the flock survives the winter.

A second supplementary feed is the compound ewe nut. It is now common practice to feed hill ewes with these nuts as lambing approaches. There are, of course, some areas where the sheep are not accessible for regular feeding, and it was thought at one time that the ewes may cease to make use of the hill grazing when a supplement was fed. This has not proved to be so, however, as the small quantity of nuts, while meeting a nutritional need, does not satisfy the appetite, and the sheep continue to search for food. To that extent, ewe nuts may be a better supplement for feeding on the hill than hay, especially when the weather is open. When snow is about, hay is still important as it is then necessary to satisfy both appetite and nutritional needs. Furthermore, there is less chance of the feed being lost under the snow.

When ewes have been brought on to the in-bye, hay is probably the better feed. While it is still highly desirable that the ewe should continue to search for her food, hay both meets her nutritional requirements and satisfies her appetite, which may help to reduce grazing when grass is precious in the late winter. It has also been produced on the farm and entails no additional expenditure in the feeding.

Animals which graze these poorer pastures for a greater part of the year will almost certainly suffer from a lack of minerals. At lambing time such a deficiency is most likely to show itself when the ewe, after several months of subsistence diet, is subjected to the stress of lambing and milk production. Many ewes are lost at this time through "staggers", which is caused by a combination of many factors, one of which is probably a mineral deficiency parallel to hypomagnesaemia in cattle. Although little is known about the exact mineral requirements of hill ewes, compound mineral supplements with a high magnesium content are often provided as an insurance both before and after lambing. Like so many other animals, hill ewes appear to sense a need for minerals at this time and take them readily. In late May and June, when ewes and lambs have returned to the hill, the improved grazing appears to satisfy their mineral requirements and any mineral supplement provided is no longer taken.

With hill sheep there is very little margin between losing sheep on the one hand through malnutrition and making a loss on the other, through over expenditure on supplementary feed; and with no other class of stock is management by the stockman, in making maximum use of limited resources, such a critical factor.

QUESTIONS—RATIONING SHEEP

1. What are the three periods in the breeding cycle when the nutrition of the ewe is critical?
2. What are the limitations on the use of feeding standards for sheep?

3. What is the purpose of increasing the nutrition of the ewe before and during tupping?
4. How is the increase in nutrition brought about and maintained?
5. What is the recommended composition of the concentrate diet at tupping time? At what rate is it fed?
6. When is it usual to start concentrate feeding prior to lambing?
7. What are the usual feeding rates for concentrates as lambing approaches?
8. To meet a ewes protein needs, what is the alternative to including high protein feeds in the diet?
9. How does the appetite of a ewe vary between mid-pregnancy, late pregnancy, early lactation, and mid-lactation?
10. What would be the approximate proportions of cereal and high protein supplement feed in a concentrate mix for ewes in the first month of lactation?
11. What would be the inclusion rate for minerals in this mix?
12. What alternative method is there for supplying minerals?
13. What factors may affect the DMI of a fattening sheep?
14. From the tables determine the NE allowance, the APL, the DMI, and the DCP allowance for the following sheep:
    (a) 25-kg lamb growing at 100 g/day;
    (b) 35-kg lamb growing at 150 g/day;
    (c) 40-kg hogg growing at 200 g/day.
15. For each of the three sheep in question 14, list the ME value and the NE values of the following feeds:

Average hay	Oats
Good quality silage	Barley
Turnips	Molassed sugar-beet pulp
Swedes	Locust bean
Rape	Groundnut cake

16. Using the feed combination chart (Fig. 4), determine the proportion of the following pairs of feeds to give the required NE content in the final ration—

	Feed A (NE)	Feed B (NE)	NE content in final ration
(a)	10.0	7.0	7.5
(b)	9.5	7.2	8.0
(c)	6.1	8.3	7.2

17. What would be the actual weights of the two feeds in each pair (kg DM) if the NE in the final ration was contained in:

    (16a)   1.25 kg DM;

    (16b)   1.10 kg DM;

    (16c)   0.95 kg DM.

18. In hill sheep farming why is it advantageous to have a number of fields or allotments of different quality grazing?

19. What are the comparative advantages of feeding hay or a ewe nut on the hill and on the in-bye as a supplementary feed?

CHAPTER 14

Rations for pigs

THERE are a number of important factors to take into account when considering pig feeding:

(a) The pig is a non-ruminant with a relatively simple digestive system and this has a number of effects on its nutrition in a number of ways:
   (i) as it does not have a rumen it is unable to digest fibre and its diet must therefore consist mainly of concentrated feeds with a minimum fibre content;
   (ii) it cannot synthesise amino acids in the same way as the ruminant and depends on a supply of the essential amino acids in its food. It is necessary, therefore, not only to consider the amount of protein in the diet but also the extent to which it supplies certain amino acids. Lysine is the amino acid which is most likely to be a limiting factor. The feeding standards for pigs state the quantity of lysine required, and the feed analysis tables give the amount in the feeds;
   (iii) again, because there is no bacterial digestion in the pig with the consequent loss of energy in the production of methane gas, nearly all the digestible energy in the feed is available for metabolism—only a little is lost in the urine. A pig's energy requirements therefore and the energy of its feed are given in terms of **digestible energy** (DE) rather than **metabolisable energy** (ME).

(b) The pig has no protective coat and is sensitive to changes in temperature, in particular to cold, damp conditions. For each pig there is a minimum temperature, the **lower critical temperature**

(LCT) at which it starts to use its feed to produce heat to counter-act the cold conditions. The LCT is related to the heat output of the pig—the higher the output the lower the LCT. The heat output of the pig depends mainly on the heat required for mainten-ance, but also on the energy used in growing. The larger the pig, therefore, and the greater its LWG, the higher is its heat output and the lower the LCT. In practice this means a suckling pig needs a house temperature of around 25°C, and a 50-kg porker about 15°C, whereas a large sow can stand a temperature as low as 5°C before it starts using its feed to keep warm. Good housing and management therefore have an important effect on **efficient feed utilisation.**

(c) Pigs are normally housed under unnatural **intensive** conditions, and depend entirely on the feeder for all their nutritional require-ments. In addition to energy and protein, particular care must also be taken to ensure an adequate supply of minerals and vitamins.

(d) Eighty per cent of the costs of pork and bacon production are feed costs and, as with no other enterprise, the profitability depends almost entirely on correct and efficient feeding.

NUTRITIONAL REQUIREMENTS

The nutritional requirements of a pig can be calculated from main-tenance and production allowances as for any other animal, but for everyday rationing of rapidly growing pigs such an approach is hardly practical. In any case there is less variation in the type of feed and environmental conditions for the pig than for other stock and pig feeding is based on a number of standard diets fed at a certain rate per day (and often *ad lib*) according to the size of the pig.

In terms of growth and development, pigs show the same pattern as other stock. In the early stages skeletal and muscular tissues develop at a greater rate than fat tissue, but at about 4–5 months of age the order is reversed and a greater proportion of fat is produced. During the early stages, diets should contain a relatively higher proportion of protein and minerals, and palatability is also important for very young pigs. The inclusion of high quality protein to ensure all the essential amino acids are present is also necessary, and for this reason the inclusion of animal protein in the form of white fish meal is highly desirable. White fish meal is also an important source of minerals, and when a diet contains 10% or more there should be no need to add a mineral supplement.

In the later stages of fattening the protein content is relatively lower and also white fish meal can be replaced by soya-bean meal as the protein source. Although this is vegetable protein it is still of high quality. When the LWG is mainly fat a high energy feed is needed. Barley is such a feed and forms the base of most pig rations, but it is particularly important in the finishing stages. It has a reputation for producing a hard white fat in the carcass, and up to 75% may be included in the finishing diet.

Vitamin supplements are essential at all ages. There are two main reasons for this: to help the pig resist disease under the intensive conditions of management, and to keep the growth rate at a maximum. Vitamins A and D should always be added, and vitamin B has also proved useful as a growth-promoting factor.

The crude-fibre content for growing pigs should not normally exceed 5%. For piglets and weaners the less fibre the better: 2–4%. Diets for adult breeding stock could contain up to 6 or 7% as an absolute maximum.

Water is essential for all pigs and it is common practice to have it always available through a continuous supply system. If the water supply should be cut off, meal intake on an *ad lib* system would immediately fall. Suckling pigs on creep feed are known to consume more when a water supply is available nearby. In general, therefore, water should be freely available for sucklers and weaners. For growing pigs 20–100 kg liveweight, 2–6 kg of water per day should be allowed, and if this is mixed with the feed the ratio of water: feed should be 2:1 by weight. (In pipeline feeding systems a ratio of 3:1 is usually used.) Pregnant sows should be allowed 4 litres/day and lactating sows from 12 to 20 litres depending on the number of piglets, although in practice a suckling sow usually has water freely available.

CONCENTRATE FEEDS USED IN PIG FEEDING

Most pig rations are based on a high percentage of home-grown or imported cereals. Barley is the most important with its high energy and low fibre content. Oats can only be used in small amounts (10–20%) mainly because of its high fibre content. Wheat is not usually available for pig feeding but it can make a valuable contribution, and up to 40% can be safely included. Maize is the main imported cereal, either flaked or as maize meal. Flaked maize is particularly useful for young pigs as it is highly palatable. Maize meal tends to give a soft yellow fat and no more than 20% is usually included.

The cereals alone do not contain sufficient protein even for pigs in the final fattening stages, and protein supplements are needed. White fish meal (580 DCP) figures in parentheses indicate g DCP/kg of the material as fed) and extracted soya-bean meal (410) together with skim milk powder (320) are high protein feeds with a high biological value. Other protein feeds with a slightly lower biological value are meat-and-bone meal (330), decorticated groundnut meal (440), and field beans (200).

Miller's offals, under the name of fine wheatfeed, weatings, or sharps is another very popular feed. It is intermediate in protein content (99) and considered to be a very safe food with a slightly laxative effect. There are various grades, however, which may vary considerably in fibre content and although the inclusion may raise the protein level of the diet it may also raise the fibre content. The better quality miller's offals, i.e. lower in fibre, are, therefore, the most valuable for pigs.

PROPRIETARY COMPOUNDS

As with other classes of stock, all the types of pig diet can be obtained as proprietary compounds. On large units, when bulk purchases can be made and bulk discounts obtained, this may be good practice. Very often such feeds are in the form of nuts or pellets, which helps to eliminate the dust from piggeries and makes the feed easy to handle and measure out.

Other farmers, who have a supply of home-grown barley and a milling and mixing installation, prefer to purchase compound protein–mineral–vitamin supplements and mix these with the barley meal. A range of diets are produced by using different supplements and barley, or in some instances different proportions of each. When this is done it is a useful exercise to check the analysis of the supplement with the merchant and satisfy oneself that the required analysis for the final diet is being attained.

Whether "straight" feeds or compound supplements are being used, cost is always an important consideration, and it may well be that precise analysis in the mix might be sacrificed in the interests of economy. Nevertheless, care has to be taken to see that the economies through the use of apparently cheaper ingredients are not lost in less-efficient feed conversion.

BASIC PIG DIETS

The nutritional requirements for pigs are met by formulating a diet corresponding to the needs of the pig at different stages of its growth. The mixes produced are either fed *ad lib* when the pig rations itself, or restricted to a ration of so many kilograms per day. It is an arbitrary decision just how many diets are formulated, but in practice five diets appear to meet the pigs' needs efficiently, on the one hand and keep the logistics of feeding feasible, on the other. Nevertheless, some farms may increase the diets to six while others manage on three. The classes of pig normally catered for are:

    (a) suckling pigs—a creep starter feed;
    (b) from 4 weeks of age to 30 kg liveweight—a weaner mix;
    (c) from 30 to 60 kg—a grower's mix;
    (d) over 60 kg—a finishing mix;
    (e) breeding stock.

The standards for these diets are given in Table 41.

TABLE 41

Standards for the basic pig diets (per kilogram as fed)

	DE	DCP	Lysine(%)	Fibre (%) (max.)
Starter creep	13.7–14.3	185–200	1.2	2
Weaner	13.5–14.0	150–160	1.0	3
Grower	12.5–13.0	135–140	0.8	4
Finisher	12.5–13.0	110–120	0.6	5
Breeding stock*	12.5–13.0	125–135	0.7	6
Pregnant sow	12.0	95–105	0.6	7

*Sometimes split between lactating sow (as shown) and pregnant sow.

STARTER CREEP

Piglets take solid feed from about 2 weeks old, and from this time on a creep feed should be available together with a supply of fresh clean water. Many breeders tempt the piglets to eat from about 7 days old by offering highly palatable foods such as milk substitute pellets or flaked maize, but these should be replaced by the normal creep by 2 weeks of age. Even so, with the small quantities involved home mixing of starter creep may not be considered worth-while. A proprietary creep with all the nutrients properly balanced, including minerals and vitamins, may cost more, but the extra cost is largely offset by a saving in labour for mixing and the guarantee of a consistent product that is easily handled.

STANDARD CREEP/WEANER

Pigs may be weaned at any time from 3 to 8 weeks and it is important that they have the weaner diet as a creep feed before weaning so that they keep eating and growing without a check. Everything should be done to encourage the piglets to eat, and a daily consumption approaching 1 kg/day at 8 weeks of age would be very satisfactory, with a total consumption of 11–12 kg up to this age. Unfortunately, this level of consumption is not always reached. The weaner mix is normally fed *ad lib* up to 30 kg liveweight.

Example A. Creep/weaner mix for pigs from 4 weeks to 30 kg liveweight (Table 42).

TABLE 42

Feed	DE		DCP		Lysine %	
	In feed	In mix	In feed	In mix	In feed	In mix
15% white fish meal	15.1	2.26	580	87	4.6	0.69
10% weatings (middlings)	11.9	1.19	99	10	0.64	0.064
20% flaked maize	15.0	3.00	70	14	0.26	0.052
55% barley	12.7	6.98	77	42	0.32	0.176
Mix analysis	—	13.43	—	153	—	0.982
Target analysis	13.5–14.0		150–160		1.0	

In this diet the high protein and mineral content is obtained by using 15% fish meal. At the same time the palatability and texture of the mix is enhanced by including 20% flaked maize. Only 10% weatings has been included, for although this feed contributes to the protein content it tends to be high in fibre. The remainder, and over half the mix, is barley meal with a relatively low fibre content and this is a cheap, palatable home-grown feed.

It is a fact that the overall energy analysis is just short of the target of 13.5, and this is because rounded proportions of the ingredients have been taken. In the following growing and finishing diets it will be seen how the ingredients have been adjusted in order to achieve the required analysis. It should also be remembered, however, that in a practical farm situation, where the feeds may need to be measured by the bag, it may not be feasible to produce a mix requiring intricate percentages of ingredients. A $2\frac{1}{2}$% unit, i.e. equivalent to one 50 kg bag in a 1 tonne mix, may be the most reliable practical mixing unit for everyday use.

Example B. A mix for growing pigs, from 30 to 60 kg liveweight (Table 43)

TABLE 43

Feed	DE		DCP		Lysine%	
	In feed	In mix	In feed	In mix	In feed	In mix
5% white fish meal	15.1	0.75	580	29	4.6	0.23
10% soya-bean meal	15.0	1.5	410	41	2.8	0.28
10% weatings	11.9	1.19	99	10	0.64	0.064
10% flaked maize	15.0	1.5	70	7	0.26	0.026
5% maize meal	14.5	1.45	73	4	0.26	0.013
60% barley	12.7	7.62	77	49	0.32	0.192
Mix analysis	13.92		140			0.805
Target analysis	12.5–13.0		135–140		0.8	

This is a cheaper mix slightly less palatable and containing a little less protein. Two-thirds of the fish meal has been replaced by soya-bean meal, the flaked maize reduced to 10%, and maize meal introduced. The home-grown cereal has also been increased.

The 60 kg liveweight coincides with the weight that many pigs are sent for slaughter for pork and up to this point the meal can be fed ad lib. There is no fear of the pigs becoming over-fat at this stage and providing the feed hoppers are working properly the amount of meal used will be no more than for restricted feeding. If it is not possible to practise *ad lib* feeding, then the amount of meal fed per day should be on the following lines:

0.5 kg of feed per day for each month of age

or 0.5 kg per day for each 12.5 kg liveweight.

If water is fed with the meal there should be an equal volume of water to meal—or approximately twice the weight of water as meal, i.e.

i.e. 2 kg meal + 4 kg (litres) of water.

Example C. A finishing mix—for pigs from 60 kg to finishing (Table 44)

TABLE 44

Feed	DE		DCP		Lysine%	
	In feed	In mix	In feed	In mix	In feed	In mix
10% soya-bean meal	15.0	15.0	410	41	2.8	0.28
10% weatings	11.9	11.9	99	10	0.64	0.064
10% maize meal	14.5	14.5	73	7	0.26	0.026
70% barley meal	12.7	88.9	77	62	0.32	0.224
Mix analysis		13.0		120		0.594
Target analysis	12.5–13.00		110–120		0.6	

The protein content has again been reduced and fish meal no longer included. Flaked maize has been replaced by maize meal and the proportion of home-grown barley again increased. A mineral supplement should now be added.

Although this is the cheapest of the diets, the pigs consume more and grow rapidly. If allowed *ad lib* feeding, although they would reach slaughter weight more quickly, they may become over-fat and grade badly. Various levels of restricted feeding are practised depending on the quality of the pigs concerned and their ability to retain the desirable conformation without becoming too fat. At 60 kg the pigs may be eating 2.1 kg of meal per day, and some feeders will restrict them to this amount. Others may allow an increase to 2.3 or 2.5 kg for pigs to be slaughtered at the 90 kg bacon weight. Pigs going to the heavy hog trade, where the amount of fat is not so critical, may well be fed over the 3-kg level.

About 5 litres of water per day is sufficient for pigs in the finishing stages.

RATIONS FOR BREEDING STOCK

The requirements for a lactating sow are greater than those of the pregnant animal, but on many farms there is only one mix for all sows although it is rationed according to the individual sow's requirements. The diet for a lactating sow approximates to that for a growing pig in terms of energy and protein content. The sow has the same need for a diet with a high protein and mineral content. Unlike the young pig, however, she is able to cope with a greater quantity of fibre in the diet, which can result in a cheaper ration.

Example D. Breeding stock, a mix for a lactating sow (Table 45)

TABLE 45

Feed	DE		DCP		Lysine%	
	In feed	In mix	In feed	In mix	In feed	In mix
10% white fish meal	15.1	1.51	580	58	4.6	0.46
20% maize meal	14.5	2.90	73	15	0.26	0.052
30% weatings	11.9	3.57	99	30	0.64	0.192
40% barley	12.7	5.08	77	31	0.32	0.128
Mix analysis		13.06		134		0.832
Target analysis	12.5–13.0		125–135		0.7	

The basic daily ration for a pregnant sow is 2.2 kg/day, but this may vary in early pregnancy, according to the condition of the sow, when she is recovering from suckling her previous litter. Late in pregnancy, as the litter is developing, she may again receive more. If the sows are at pasture the basic ration should be modified to reflect the quantity and quality of the grazing, but throughout it must be the condition of the sow which is the governing factor.

The same rule must apply to the suckling sow, but as an approximate guide the daily ration from 1 or 2 weeks after farrowing is as follows:

Basic ration, 3 kg;
Production ration, 0.25 kg for each piglet.

These figures are only a guide and a sow who is obviously milking well can be fed up to 6–6.5 kg/day. Towards the end of the suckling period, if the sow is in reasonable condition and the piglets are taking the creep feed readily, the ration may be reduced. A good supply of clean water should always be available.

If a separate diet is prepared for pregnant sows, then this could be of lower nutritive value approximating more closely to a fattening ration.

Example E. Breeding stock, a mix for a pregnant sow (Table 46)

TABLE 46

Feed	DE		DCP		Lysine %	
	In feed	In mix	In feed	In mix	In feed	In mix
5 % white fish meal	15.1	0.75	580	29	4.6	0.23
25 % weatings	11.9	2.98	99	25	0.64	0.16
70 % barley	12.7	8.89	77	54	0.32	0.22
Mix analysis	—	12.62	—	108	—	0.61
Target analysis		12.00		95		0.6

FEED ADDITIVES

It has become common practice in recent years to add small amounts of certain substances to the feed which have been shown to improve growth rate and the efficiency of feed conversion, particularly in intensively housed stock. Some of these are antibiotics or other substances with an antibacterial action, but copper has for some time been one of the most common. The action of some of these materials is not fully understood and the active ingredient of some proprietary compounds is kept a close secret; but the majority would seem to have an effect on the bacterial content of the digestive system which has a beneficial effect on the animal.

It would seem, therefore, that they tend to prevent or control disease and they have been shown to have been most beneficial under poor conditions of housing and management when the disease risk is high. Nevertheless, response in trials has been very variable and sometimes negative. It is not easy, therefore, to make recommendations on which additive to use except that copper has been proved and is cheap. Any other additive should be considered relative to copper.

As such small quantities are used (for copper it is 175–200 ppm), it is essential that the instructions for proprietary compounds are followed closely. It is usually necessary to make a pre-mix of the ingredient with a quantity of meal, which is then added to the main mix, in order to make sure the substance is evenly spread through the feed.

As a general rule it is wise to consult a veterinary surgeon and have his approval of any feed additive that is to be used. It is, of course, illegal to feed antibiotics continuously in pig feeds. The other rule that must be followed is never to feed the meal which contains an additive to any other stock than those for which it was prepared.

MINERALS AND VITAMINS

The importance of white fish meal as a source of minerals has already been mentioned. When this feed is not included in a diet there is a need to supplement the ration with a mineral mix usually at a rate of 1–2%. Calcium carbonate, di-calcium phosphate, and common salt can be added cheaply, but a proprietary mix will provide these major minerals together with a number of trace elements for little additional cost.

In a similar way a vitamin supplement should also be added to the diets as a matter of form. Cod-liver oil supplies vitamins A and D_3 and liquid vitamin supplements usually have cod-liver oil as a base. Vitamins can also be obtained in a crystalline form.

PIG FEEDING AND PIG MANAGEMENT

The aim in this chapter has been to demonstrate how diets and rations can be formulated from the basic nutritional needs of the stock and from a knowledge of the feeds available. In pig feeding, however, the system of management, and particularly the method of housing, has a major influence, and it is not really satisfactory to consider any one aspect in isolation. Because of other management factors, meal may be fed wet or dry, as meal crumbs or larger pellets, *ad lib* or restricted. The decision to change from *ad lib* to restricted feeding often coincides with a change of housing or some other management factor. The quality of the housing itself clearly has an influence on the efficiency with which a pig converts its feed into liveweight gain.

A source of cheap feed, often some form of catering waste, may be available, and this may dominate the whole system of feeding and management on a particular holding.

TABLE 47

Feeds commonly used in pig diets

The nutritive values given are for the feed *as fed*

	Digestible energy (DE) (MJ DE/kg)	Digestible crude protein (DCP) (g/kg)	Lysine (g/kg)	Crude fibre (%)
Cereals—high energy				
Barley	12.7	77	3.2	4.6
Maize	14.5	73	2.6	2.0
Oats	11.4	78	3.7	10.4
Wheat	14.0	86	2.8	2.2
Flaked maize	15.0	70	3.6	1.5
Medium protein				
Wheatings (fine middlings)	11.9	99	6.4	7.5
Field beans	12.6	200	16.0	7.3
High protein (vegetable)				
Groundnut meal (dec.)	16.1	442	18.0	7.9
Soya bean meal	15.0	410	28.0	5.2
High protein (animal)				
Dried skimmed milk	16.5	326	22.0	—
Meal and bone meal	7.8	359	33.0	—
White fish meal	15.1	580	46.0	—

Finally, of course, there is the all-important factor of the market that is being catered for, which may be anything from quality weaners to heavy hogs; this, too, can have a radical influence on the feeding system which is practised.

QUESTIONS

1. A pig is a non-ruminant—what effect on its feeding does this have?
2. What do you understand by the term lower critical temperature?
3. What is the LCT for (a) a suckling pig (b) a 50-kg porker?
4. What percentage of costs are feed costs in pig production?
5. What is the importance of white fish meal as a pig feed?
6. What is the main source of vegetable protein in pig feeding?
7. What vitamins are added to pig rations?
8. What is the maximum quantity of crude fibre allowable in pig rations?
9. What is the usual water:feed (by weight) ratio if water is being added to the meal of fattening pigs?
10. For what five classes (state liveweight where applicable) of pigs are diets produced?
11. What is the normal range of digestible energy in these diets?
12. How does the protein level vary over the range of diets?
13. Name two highly palatable feeds that might be used in a starter creep diet.
14. In compiling diets why is a $2\frac{1}{2}\%$ unit significant?
15. If growing pigs are not fed *ad lib*, what would be the relationship between meal fed/per day and age?
16. At what weight might restricted feeding start?
17. To what weight of feed would the pigs be restricted?
18. To what extent is a diet for a sow similar and different from that of a young growing pig?
19. What is the basic daily ration for an in-pig sow?
20. How may this be varied in pregnancy?
21. What weight of meal per day would be fed to a sow with ten piglets?
22. Why are feed additives included in diets?

23. How are they thought to act?
24. Name a common feed additive.
25. What system of mixing should be practised with additives?
26. What is the normal inclusion rate for a mineral supplement?
27. What are the vitamins in cod-liver oil?

Index

Abdominal cavity 41
Abomasum 41, 45
Absorption 39
Accessory organs 39
Adipose tissue 21
Alimentary canal 38–49
Amides 78
Amino acids 24, 26, 35
Anaemia 31
Animal production level (APL) 105,
    108, 109, 111, 129
Anus 42
Appendix 42
Appetite 88, 102
    See also DMI

Beef 102–16
    appetite—DMI 102
    APL 105, 108, 109, 111
    barley beef production 66, 113–14
    energy
        for maintenance 103
        for production 104
    finishing cattle – silage/cereal 110–11
    growing cattle
        ME allowances 106
        NE allowances 108
        NE value of feeds 109
        rations 107
    M/D 104
    minerals 116
    protein requirements 109
    vitamins 116
Biological value 26, 59
Blood capillaries 42

Bran 64
Brewers' grains 65

Cabbage 55
Caecum (blind gut) 42, 45
Calcium 30, 31, 32
Carbohydrates 7, 13–17
    chemistry of 13
    in the animal 15
    in the food 14
Carbon 13
Carnivorous 40
Cellulose 14, 15
Chalk 32
Chlorine 30, 32
Chlorophyll 14
Coarse fodders (roughages) 47
Cod liver oil 36
Colon 42
Common salt 32
Compound feeds 59, 67
Concentrates 58–68
Cotton cake 62
Creep feed (pigs) 143
Cudding 46

D-value 98–9
Dairy cattle
    rationing 82–100
    appetite—DMI 88–9
    compiling a ration 89
    DCP allowances 87
    fibre content 93, 99
    liveweight change 85

153

Dairy Cattle—*contd.*
　maintenance　84
　ME allowances　87
　minerals　96
　production　84
　summer feeding　98
　vitamins　97
　winter feeding　82
Diabetic　16
Diets　74
Digestibility　27, 79
Digestion　38–47
Digestive systems
　horses　44
　pigs　43
　ruminants　44
Digestive tract　38
Disaccharides　13
Distillers' grains　63
Dried grass　64
Dry matter (DM)　7, 11, 73
Dry matter intake (DMI)　73, 88–9, 102, 129
Duodenum　41

Egestion　39
Energy　75–7
　balancing chart　91
　digestible (DE)　75–7, 138
　gross (GE)　75–7
　metabolisable (ME)　75–7, 82, 129
　net (NE)　76–7, 108–9, 129
　values　76–7
Enzyme　38
Epiglottis　40
Ether extract　20

Faeces　42
Fats and oils　7, 19–22
　chemistry of　19
　in the animal　21
　in the plant and feeds　20
　marbling fat　21
　reserve fat　21
　tissue fat　21
Fatty acids　19–20
Feed additives　147

Feed analysis tables　68
Feed combination chart　130
Feeding standards　80, 119
Feeds commonly used　49–69
Fibre　14, 15, 93
Flaked maize　65
Fodder beet　53
Food　4–5
　constituents　7
　functions of　4–5

Gastric juice　41
Glucose　15
Glycerol　19
Grain balancer cake　59
Grass　56
Grass staggers　56
Green fodder　49, 53–5
Groundnut cake/meal　62

Hay　50
Heat increment　76
Herbivorous　40
High protein compounds　59
Hill sheep　134
Horse　43–4
Hydrochloric acid　41
Hydrogen　13
Hypomagnesaemia　56

Intestine　39, 41, 42
Iron　31

Kales　54

Lactose　14
Lignin　14–15
Linseed cake　62
Liver　39, 41
Locust beans　66
Lower critical temperature (LCT)　138
Lymph vessels　42
Lysine　138

M-over-D (M/D)   78, 104
Magnesium   29
Maintenance   6, 74, 84, 103, 129
Maize   65
Maize germ meal   66
Maltose   14
Mangolds   52
Meat and bone meal   60
Megajoule (MJ)   76
Mesenteric membrane   39
Metabolism   31, 34, 38
Milk fever   31
Minerals   7, 29–33
    beef   116
    dairy cows   96
    deficiency   30–1
    pigs   148
    sheep   125
    supplements   32
Monosaccharides   13
Mouth   39
Mucin   40

Nitrogen   23
Nitrogen free extractives   18

Oats   67
Oesophageal groove   45
Oesophagus (gullet)   40–1
Omasum   45
Omnivorous   40
Organic catalyst   34
Oxygen   13, 16

Palatability   119, 141
Palate, soft and hard   39
Palm kernel   63
Pancreas   39, 41
Pea and bean meal   63
Pepsin   41
Pharynx   39
Phosphorus   23
Photosynthesis   16
Pigs
    rations for   138–51
    basic diets   142

breeding stock   146
concentrate feeds   139
creep   143
DE and ME   138
feed additives   147
finishing   145
growing   144
lower critical temperature (LCT)   138
minerals   148
nutritional requirements   139
proprietary compounds   141
weaner   144
vitamins   139, 148
Polysaccharides   14
Potatoes   53
Production   6, 74, 84, 103
Protein   7, 23–8
    chemistry of   23
    crude protein   24, 78
    digestible crude protein   27, 28, 79
    degradable/undegradable   86
    in feeds   24
    in the animal   25
    measurement of   78
Protein balancing chart   92
Ptyalin   40

Rape   55
Ration   74
Rectum   42
Rennin   41
Respiration   16
Reticulum   45, 46, 49
Rumen   44, 45, 47
Ruminants   44–5

Saliva   40
Salivary glands   40
Separated milk   60
Sheep
    rations for   118–35
    APL table   126
    breeding sheep allowances   115
    critical periods   118
    feed combination chart   130

Sheep—*contd.*
  feeding standards 119
  finishing hoggs 128
  hill sheep 134
  lactation 123
  metabolisable energy (ME) allowances 127, 129
  minerals 125
  net energy (NE) 126, 129
  NE value of feeds 126
  palatability 119
  pregnancy 120
  protein allowances 128
  tupping 120
  vitamins 133
Silage 56–8
Solanin 53
Soya bean cake/meal 61
Sphincter muscles 41
Starch 14
Steamed bone flour 32
Stomach 39, 41
Straw 51
Succulent feeds 52
Sucrose 14

Sugar beet
  pulp 66
  tops 55
Summer feeding (dairy cattle) 198
Swedes 52
Swede tops 55

Teeth 39
Turnips 52

Villi 42
Vitamins 7, 34–6
  beef 116
  dairy cows 97
  pigs 139, 140, 148
  sheep 133
  supplements 36

Water 9–11
  in the animal 7, 9, 10
  in the food 10, 11
Weatings 64
Wheat 65
White fish meal 59